AF342523

NEW DEVELOPMENTS IN CALCIUM SIGNALING RESEARCH

NEW DEVELOPMENTS IN MEDICAL RESEARCH

Additional books in this series can be found on Nova's website
under the Series tab.

Additional e-books in this series can be found on Nova's website
under the e-book tab.

NEW DEVELOPMENTS IN CALCIUM SIGNALING RESEARCH

MASAYOSHI YAMAGUCHI
EDITOR

Nova Biomedical

New York

For permission to use material from this book please contact us:
Telephone 631-231-7269; Fax 631-231-8175
Web Site: http://www.novapublishers.com

NOTICE TO THE READER

The Publisher has taken reasonable care in the preparation of this book, but makes no expressed or implied warranty of any kind and assumes no responsibility for any errors or omissions. No liability is assumed for incidental or consequential damages in connection with or arising out of information contained in this book. The Publisher shall not be liable for any special, consequential, or exemplary damages resulting, in whole or in part, from the readers' use of, or reliance upon, this material. Any parts of this book based on government reports are so indicated and copyright is claimed for those parts to the extent applicable to compilations of such works.

Independent verification should be sought for any data, advice or recommendations contained in this book. In addition, no responsibility is assumed by the publisher for any injury and/or damage to persons or property arising from any methods, products, instructions, ideas or otherwise contained in this publication.

This publication is designed to provide accurate and authoritative information with regard to the subject matter covered herein. It is sold with the clear understanding that the Publisher is not engaged in rendering legal or any other professional services. If legal or any other expert assistance is required, the services of a competent person should be sought. FROM A DECLARATION OF PARTICIPANTS JOINTLY ADOPTED BY A COMMITTEE OF THE AMERICAN BAR ASSOCIATION AND A COMMITTEE OF PUBLISHERS.

Additional color graphics may be available in the e-book version of this book.

Library of Congress Cataloging-in-Publication Data

ISBN: 978-1-62948-601-7

Library of Congress Control Number: 2013954141

Published by Nova Science Publishers, Inc. † *New York*

Contents

Preface

Calcium metabolism in living body is skillfully regulated through calcium-regulating hormones including parathyroid hormone, active vitamin D_3 metabolite and calcitonin. Blood calcium concentration is ten thousand-fold as compared with that in the cells. Cells use excellently extracellular calcium to regulate cell function. Intracellular calcium homeostasis is exactly regulated through mechanism related to various calcium transport systems in cells. Signal transduction plays a pivotal role in cellular regulation. The second messenger is generated in cells once the first messenger binds to the receptors of plasma membranes. The second messenger, which was found firstly, is cyclic adenosine monophosphate. After that, intracellular calcium has been demonstrated to play a role as a second messenger for hormonal stimulation in cells. Calcium signaling is probably the most ubiquitous cellular signal that mediates the action of many hormones, cytokines, and neurotransmitters. Calcium regulates the activity of numerous calcium-dependent proteins including protein kinases, protein phosphatases, protease and other calcium-regulating factors, which are involved in calcium signaling. Calcium signaling plays a pivotal role in the regulation of manifold cellular functions including cell proliferation, apoptosis and gene expression. This book focuses recent topics in calcium signaling and its related diseases.

Chapter 1: This chapter discusses the evidence for a novel receptor, vitamin D metabolite, 1, 25-dihydroxyvitamin D_3. The protein, which author have termed as the membrane-associated, rapid response steroid–binding (1,25D_3-MARRS) receptor, is identical to ERp57/PDIA3. This protein has been extensively studied for its role in calcium and phosphate uptake in intestinal cells and more recently has been found to be implicated in cancer and Alzheimer's disease. In addition, author present a more complete biochemical characterisation of this protein using co-immunoprecipitation and chromatin immunoprecipitation studies. The 1,25D$_3$-MARRS protein has both cytoplasmic and nuclear motifs and redistribution within the cell. New evidence shows its potential involvement through genomic action. Its involvement in disease states appears primarily through its chaperone action, post-translational modification of vital proteins and ability to modulate oxidative stress.

Chapter 2: Cystic Fibrosis (CF) is characterised by ion transport abnormalities caused by a mutation of the Cystic Fibrosis Transmembrane conductance Regulator (CFTR) gene. This results in airway surface dehydration, impaired muco-ciliary clearance, favours chronic bacterial infection and sustained inflammation and leads to progressive lung destruction. In CF, intracellular calcium status represents a double-edged sword. Indeed, increases in intracellular calcium can restore the airway surface liquid layer architecture and muco-ciliary clearance via stimulation of the calcium-activated chloride currents. However, heightened intracellular calcium activity has been implicated as a major participant in promoting hyper-inflammatory pathways via its role in endoplasmic reticulum biology and its second messenger function, interacting with calcium-sensitive proteins and promoting nuclear transcription of pro-inflammatory mediators. In this chapter, the authors explore how glucocorticoids and lipoxins differentially regulate intracellular calcium concentration, airway hydration and inflammation in CF lung disease. In particular, the authors highlight how Lipoxin A$_4$ (LXA$_4$) couples the beneficial effects of raised intracellular calcium concentration to promote chloride secretion and airway hydration to an anti-inflammatory and pro-resolution programme of transcription.

Chapter 3: The function of neurons, cardiomyocytes and other excitable cells is affected by hormones and the changing patterns of the plasma membrane electrical activity. Voltage-gated Ca$_v$1.2 calcium channels and cAMP/G protein-coupled receptors are the two major pathways communicating external excitation to the cell nucleus to regulate transcription in the process named excitation–transcription coupling. The exact signaling events involved in excitation–transcription coupling are poorly understood. In this chapter the author discuss the recent progress made in the research of transcription regulation by the Ca$_v$1.2 calcium channels. The author will focus on the use of wavelet transform, the new powerful statistical approach to FRET image analysis, which revealed the spatio-temporal domain organization of the nuclear signaling associated with excitation–transcription coupling.

Chapter 4: Not only calcium transient but also repetitive calcium elevation plays an important role for the elucidation of cellular responses for various types of stimulations. The pattern of calcium oscillation can be classified into three groups according to frequency; normal (less than 1 Hz), slower (milli Hz) and ultra-low (micro Hz). These three groups of calcium oscillation are associated with different values of temperature dependent coefficient, Q10. The Q10 value for normal frequency was found to be about 4, which is corresponding to ATP consumption process. The value of Q10 = 2 is well known for general biochemical reactions, while Q10 = 1 is matching to physical process; ionic diffusion. Normal frequency with Q10 = 4 is valid for keeping secretion constant for long time. On the other hand, ordinary slower oscillation with Q10 = 2 means that the calcium oscillation is guaranteed by enzymatic reactions. The rate determinant process of ultra-low frequency oscillation is due to ionic diffusion (Q10 = 1.0), where diffusion coefficient of ions is exceptionally reduced due to bound water on the surface of proteins. Although intracellular water structure is highly dependent on the cell cycle; bulk (free) water in G1 and G2, and bound water in S and M phase. Since period of G1 and G2 occupied 60 %~70 % of full period of one cell cycle, which means about 70 % of cells in the in vitro experiment are staying at G1 and G2 phase in average. In the case of circadian rhythm experiment, cell cycle is arrested, thus the obstacle of ionic diffusion by bound water must be considered. Eventually a role of calcium oscillation may be a kind of cell clock in cell cycle and circadian rhythm, while time interval calcium elevation is needed for secretion process to avoid cell death.

Chapter 5: The common liver fluke, *Fasciola hepatica*, is a parasite of mammals. In the western world its effects are largely felt on agriculture where infection of cows, sheep and other farm animals is estimated to cause millions of dollars of financial losses. In the developing world, the problem is even more serious with an estimated 7 million infected people and many millions more at risk of infection. Calcium signalling is of key importance in all eukaryotic species and recent discoveries of novel types of calcium binding proteins in liver flukes (and related trematodes) suggest that there may be calcium signalling processes which are unique to this group of organisms. If so, these pathways may provide potential targets for the design of novel anthelmintic drugs. Here, author review three main groups of *F. hepatica* calcium binding proteins: the FH8 family, the calmodulin family (FhCaM1, FhCaM2 and FhCaM3) and the EF-hand/dynein light chain family (FH22, FhCaBP3, FhCaBP4). Considerable information has been gathered on the sequences, predicted structures and biochemical properties of these molecules. The challenge now is to understand their functions in the organism.

Chapter 6: Regucalcin is discovered in 1978 as a calcium-binding protein that does not contain EF-hand motif of calcium-binding domain, which differs from calmodulin and other calcium-related proteins. The name, regucalcin, was proposed for this calcium-binding protein, which regulates various calcium- or calcium/calmodulin-dependent enzyme activations. Regucalcin and its gene (*rgn*) are identified in over 15 species consisting of regucalcin family. The regucalcin gene is localised on X chromosome. Regucalcin is greatly pronounced in hepatocytes and kidney proximal tubular cells. This chapter provides recent information concerning the role of regucalcin in the regulation of kidney cell function. Regucalcin has been demonstrated to play a physiological role in the regulation of calcium homeostasis in the kidney proximal tubular epithelial cells, which participate in transcellular transport to reabsorption of calcium from filtrated urinary calcium. Regucalcin has also been shown to regulate various enzyme activities including calcium-dependent protein kinases and protein phosphatases, nitric oxide synthase, cyclic adenosine monophosphate phosphodiesterase, and others, which are involved in intracellular signaling pathways. Regucalcin is localized to the nucleus and regulates the gene expressions of various proteins, which are related to mineral transport-related proteins, Smad 2, the nuclear factor-kappa B, cell proliferation and apoptosis-related proteins in the kidney tubular epithelial cells. Regucalcin reveals suppressive effects on cell proliferation and apoptotic cell death that are induced through various signaling stimulators including calcium. In this chapter, regucalcin is introduced to play a pivotal role as a multifunctional protein in the regulation of kidney proximal tubular epithelial cells.

Chapter 7: Regucalcin has been demonstrated to play a multifunctional role in cell regulation including the depression of apoptotic cell death and cell proliferation induced by various signaling factors in many cell types *in vitro*. Cytoplasm regucalcin translocates to nucleus and suppresses nuclear deoxyribonucleic acid (DNA) and ribonucleic acid (RNA) synthesis. Calcium has been shown to regulate protein, DNA and RNA synthesis. In this chapter, regucalcin has been shown to play suppressive roles in protein synthesis and degradation in relation to calcium regulation. Regucalcin has been found to suppress aminoacyl-tRNA synthetase and activate protease in cells. Regucalcin may play a suppressive role in protein output in cells.

The editor believes that this book, which has been written with the recent topics of calcium signaling and its related diseases, will be of interest to scientists focused in the research fields of molecular and cellular biology and biomedicine concerning calcium signaling.

In: New Developments in Calcium …   ISBN: 978-1-62948-601-7
Editor: Masayoshi Yamaguchi   © 2014 Nova Science Publishers, Inc.

Chapter I

The 1,25 Dihydroxyvitamin D₃-Membrane–Associated, Rapid Response Steroid-Binding Receptor: Physiological Roles and Characterization of Binding Partners

T. M. Sterling[1], R. C. Khanal[1,2] and I. Nemere[1,]*

[1]Department of Nutrition, Dietetics and Food Science,
Utah State University, Logan, Utah, US
[2]Global Future Institute, Dallas, Texas, US

Abstract

In this chapter, we discuss the evidence for a novel receptor for the vitamin D metabolite, 1,25-dihydroxyvitamin D_3. The protein, which we have termed the membrane-associated, rapid response steroid–binding (1,25D_3-MARRS) receptor, is identical to ERp57/PDIA3. This protein has been extensively studied for its role in calcium and phosphate uptake

---

[*]Corresponding author: I. Nemere, E-mail: ilka.nemere@usu.edu.

in intestinal cells and more recently has been found to be implicated in cancer and Alzheimer's disease. In addition, we present a more complete biochemical characterisation of this protein using co-immunoprecipitation and chromatin immunoprecipitation studies. The $1,25D_3$-MARRS protein has both cytoplasmic and nuclear motifs that contribute to redistribution within the cell. New evidence shows its potential involvement in genomic action. Its involvement in disease states appears primarily through its chaperone action, post-translational modification of vital proteins, and ability to modulate oxidative stress.

Abbreviations

$25D_3$	25-hydroxyvitamin D_3
$1,25D_3$	1,25-dihydroxyvitamin D_3
$1,25D_3$-MARRS	membrane-associated, rapid response steroid-binding
cAMP	cyclic adenosine monophosphate
CHML	choroidermia-like gene
co-IP	co-immunoprecipitation
Erp57	endoplasmic reticulum p57
GPR180	G protein-coupled receptor 180
HOMER1	Homer homologue 1
JNK	c-Jun N-terminal kinase
LAMP3	lysosomal-associated membrane protein 3
MAPK	mitogen-activated protein kinase
MCF-7	Michigan Cancer Foundation-7
MISS	membrane-initiated steroid signalling
NEURL1B	neutralized-like protein 1B
NISS	nuclear-initiated steroid signalling
nVDR	nuclear vitamin D receptor
PDIA3	protein disulphide isomerase-3
PKA	protein kinase A
PKC	protein kinase C
SAPK	stress-activated protein kinase
SLC30A1	zinc transporter 1
STRING	Search Tool for the Retrieval of Interacting Genes/ Proteins
TBC1D8	TBC1 domain 8
UVR	ultraviolet radiation
VDR	vitamin D receptor.

Introduction

The field of steroid research has evolved significantly during the past 60 years focussing on the concept that lipophilic signalling agents lack membrane receptors and act solely through gene transcription. It is now widely accepted that steroid hormones have membrane receptors and can elicit responses that are rapid or 'pre-genomic'.

Some of these receptors are the classical nuclear transcription factors that have been modified with a fatty acyl chain (e.g. palmitoylation) to promote association with the plasma membrane. However, in several cases of steroid research, novel receptors have been identified. This review will describe one such novel receptor, the membrane-associated, rapid response steroid-binding (1,25D$_3$-MARRS) receptor for the steroid hormone 1,25-dihydroxyvitamin D$_3$ (1,25D$_3$).

Discussion

In this review, the authors have referenced some of their own studies.

These referenced studies have been conducted in accordance with the Declaration of Helsinki (1964) and the protocols of these studies have been approved by the relevant ethics committees associated to the institution in which they were performed.

All human subjects, in these referenced studies, gave informed consent to participate in the studies. Also, animal care was in accordance with the institutional guidelines.

Figure 1 presents a schematic presentation of the metabolism of vitamin D to 25-hydroxyvitamin D$_3$ (25D$_3$) in the liver.

This metabolite is hydroxylated in the kidney either to 1.25D$_3$ when an animal is calcium, phosphate or 1,25D$_3$ deficient or to 24,25D$_3$ in a replete animal. 1,25D$_3$ is best known for its stimulatory activity, for example, it increases calcium and phosphate absorption in the intestine [1, 2], while 24,25 D$_3$ is known for its inhibitory activity and can cause hypophosphataemia *in vivo* [3].

Major Metabolites of Vitamin D

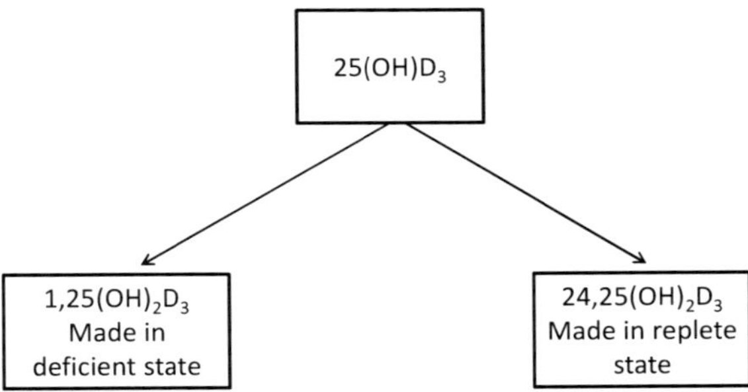

Figure 1. Schematic representation of the major metabolites of vitamin D. Vitamin D is hydroxylated at C-25 in the liver and then either metabolised to 1,25D$_3$ in the kidney or 24,25D$_3$. 1,25D$_3$ is a metabolite that stimulates calcium and phosphate absorption, and hence is produced during deficient states. In the replete states, 24,25D$_3$ is produced and it acts as an inhibitor of the stimulatory action of 1,25D$_3$.

The 1,25D$_3$-MARRS Receptor: Rapid Responses for Calcium and Phosphate Uptake

1,25D$_3$ has been shown to stimulate calcium uptake in intestinal cells from rats [4], chicks [5] and mice [1]. Each of these has been shown to require the 1,25D$_3$-MARRS receptor, most recently shown in mice that were genetically engineered to have a targeted knockout of the gene in their intestine [1].

Similarly, phosphate uptake has been shown to depend on the functional presence of the 1,25D$_3$-MARRS receptor in these three species [2, 6, 7].

In 2004, our laboratory published results [6] using chick intestinal cells and two independent techniques to study the requirement of the 1,25D$_3$-MARRS receptor in hormone-stimulated phosphate uptake.

Using a highly specific polyclonal antibody reagent to the N-terminus (generated by the multiple antigenic peptide procedure), we pre-incubated the intestinal cells with this reagent and subsequently demonstrated that while 1,25D$_3$ treated cells exhibited a significant increase in phosphate uptake relative to controls, those treated with a 1/500 dilution of Ab 099, created against 1,25D$_3$-MARRS, failed to respond to hormone with increased phosphate uptake. The independent techniques used cells transfected with either a ribozyme directed against the 1,25D$_3$-MARRS receptor or the control ribozyme. Those cells transfected with the ribozyme against the receptor were found to have decreased protein levels as demonstrated by the Western blot analysis and exhibited a failed response to 1,25D$_3$ with increased phosphate uptake relative to the controls and cells treated with the control ribozyme. Protein kinase C (PKC), which mediated the phosphate uptake response in chick intestinal cells, also showed a failed response to stimulation by hormone in the cells transfected with active ribozyme.

Finally, we demonstrated a decreased [^{3}H]1,25D$_3$ binding to the 1,25D$_3$-MARRS receptor in cells transfected with the active ribozyme, while the classical vitamin D receptor (VDR) was found to be unaffected in its [^{3}H]1,25D$_3$ binding.

More recently, we used mice bearing a targeted knockout of the 1,25D$_3$-MARRS receptor in intestine that was produced by crossing mice bearing the floxed intestinal cell gene for the receptor with mice bearing villin-driven cre-recombinase. For calcium uptake studies, we found that this protein was required for hormone-stimulated uptake. We also discovered that as in chick intestinal cells [5], the protein kinase A (PKA) pathway also mediated the calcium uptake [1].

The 1,25D$_3$-MARRS receptor was similarly required for the 1,25D$_3$-mediated phosphate uptake [2]; however, the PKC pathway was not involved. We also tested whether the 1,25 D$_3$-MARRS receptor knockout had observable effects on calcium absorption *in vivo*, where we administered ^{45}Ca Cl$_2$ by oral gavage, followed by a 5 min absorption and collection of blood by cardiac puncture. The knockout mice had less calcium absorption when compared with the littermates as determined by the serum ^{45}Ca [8].

Receptor Isolation and Characterisation

The need for a membrane receptor to mediate the rapid or 'pre-genomic' effects of the steroid hormone 1,25D$_3$, was postulated by Nemere and Szego [4] when they observed rapid (within 5 min) 1,25D$_3$-stimulated calcium uptake commensurate with lysosomal enzyme release in rat enterocytes. Since then, several investigators have demonstrated that 1,25D$_3$ stimulates changes in phospholipid metabolism within the brush border membranes of chick [9, 10] and rat [11] enterocytes that are independent of 'genomic' effects as well as $1,25D_3$-stimulated calcium uptake that is independent of transcription or translation events in mouse mammary glands [12] and chick duodena [13]. $1,25D_3$ also stimulates signalling events in rat enterocytes [14] and porcine parathyroid cells [15] specific to the plasma membrane that induces phospholipid metabolism to increase concentrations of chemical messengers, such as inositol triphosphate, diacylglycerol, phosphatidic acid and/or monoacylglycerol within seconds. These representative examples of non-nuclear or pre-genomic actions, along with the identification of a protein that specifically binds $1,25D_3$ in the basal lateral membrane of chick enterocytes to facilitate the calcium uptake [16, 17], has led to the identification of a plasma membrane receptor, now termed the $1,25D_3$-MARRS receptor.

As a novel receptor for the steroid hormone $1,25D_3$, the $1,25D_3$-MARRS receptor has been identified through sequence analysis as a protein that is distinct from both the nuclear VDR (nVDR) and the vitamin D binding protein [18]. It was first identified as a 66 kD membrane protein found at the basal lateral membrane of chick duodena that specifically binds the $1,25D_3$ in a manner that is not only saturable [16, 17, 19–22], but also conserved in the plasma membranes of cells and vesicles across several species [6, 23], including mouse, rat, human and fish [19, 24] and is found in several tissues, such as the intestine, bone, teeth, parathyroids, muscle, kidneys and the brain [25–32]. The $1,25D_3$-MARRS receptor has been shown to facilitate various cell signalling events, such as $1,25D_3$-stimulated activation of the PKC pathway, cyclic adenosine monophosphate (cAMP) pathway, mitogen-activated protein kinase (MAPK) pathway and endocytic/lysosomal transport pathway.

The $1,25D_3$-MARRS receptor contains several consensus sequences that shed light on the membrane-initiated steroid signalling (MISS) events that are involved in 'pre-genomic' cellular effects that include an N-myristoylation site, seven PKC sites, a cAMP-dependent kinase site, two thioredoxin folds, six casein kinase II sites and three tyrosine kinase sites.

The N-myristoylation site of the $1,25D_3$-MARRS receptor acts as a signal that targets the receptor for insertion into the plasma membrane. The binding of $1,25D_3$ to the $1,25D_3$-MARRS receptor at the plasma membrane, is shown to act through a G-protein (G_{aq}) [33] and the PKC pathway [34–36] or the cAMP pathway [21, 22, 28] to initiate signal transduction events that enhance calcium uptake and/ or transport [21, 22], calcium extrusion [28] or phosphate uptake and/or transport [6, 32, 36, 37].

The 1,25D₃-MARRS receptor also contains two thioredoxin folds that facilitate its ability to interact with calcium binding proteins, such as calnexin and calreticulin to modulate intracellular calcium concentrations (Figure 2). The binding of 1,25D₃ to the 1,25D₃-MARRS receptor can either activate the MAPK pathway through tyrosine kinase phosphorylation to elicit responses that are also PKC-dependent [33, 38–41], or possibly through the Wnt signalling pathway through its casein kinase II site activity.

While these characteristics of the 1,25D₃-MARRS receptor clearly establish its involvement in 1,25D₃–stimulated 'pre-genomic' events, the 1,25D₃-MARRS receptor also contains consensus sequences such as two Rel homology domains, a SMAD3 consensus sequence, a nuclear retention motif and an endoplasmic reticulum retention motif [6, 18, 23, 42], which demonstrates that the 1,25D₃-MARRS receptor is also integral in 1,25D₃-stimulated cell signalling that involves the endocytic/lysosomal transport pathway to mediate cell events that can integrate 'pre-genomic' signalling events with longer-term genomic responses [22–24].

1,25(OH)₂D₃ Predicted to Bind Near C31-34 in Thioredoxin Fold (a)

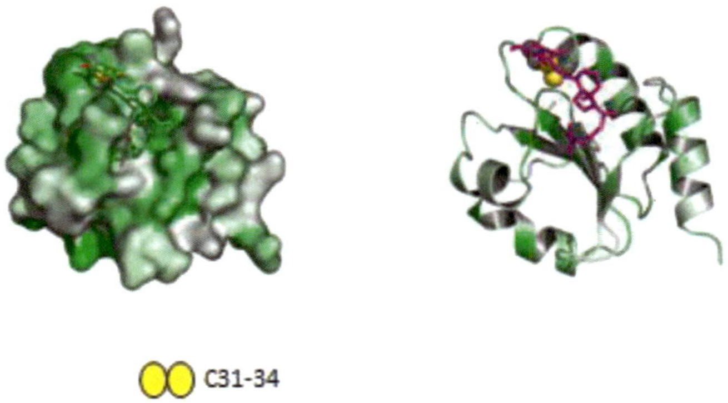

Figure 2. Two models of the 1,25D₃-MARRS receptor. 1,25D₃ is predicted to bind near the N-terminus as indicated in both the space filling and ribbon models.

Chromatin Immunoprecipitation and Co-Immunoprecipitation Studies

While MISS events may occur independently of nuclear-initiated steroid signalling (NISS), the $1,25D_3$-MARRS receptor is able to bind to DNA through its Rel homology domain interaction sites [6] (as well as a SMAD3 consensus sequence [23, 42]) and has been shown to translocate to the nucleus [43, 44].

Using the chromatin immunoprecipitation studies, we have recently described the $1,25D_3$-MARRS receptor interactions with several genomic sequences both prior to $1,25D_3$ exposure in murine enterocytes and within 5 min of hormone addition (see Table 1).

Prior to $1,25D_3$ exposure, the $1,25D_3$-MARRS receptor interacts with the Homer homologue 1 *(HOMER1)* gene and the protein disulphide isomerase-3 *(PDIA3)* gene, which facilitate changes in intracellular calcium.

While *HOMER1* helps regulate intracellular calcium by coupling surface receptors, such as glutamate, IP3 ryanodine receptors, or cytoplasmic proteins such as PI3 kinase, PDIA3 acts as a chaperone protein to mediate calcium sequestration by modulating the activity of calnexin and calreticulin. After the $1,25D_3$ exposure, the $1,25D_3$-MARRS receptor interacts with the neutralised-like protein 1B *(NEURL1B)* gene, zinc transporter 1 *(SLC30A1)* gene and/or the G protein-coupled receptor 180 *(GPR180)* gene. *NEURL1B* encodes the sequence of E3 ubiquitin protein ligase that targets proteins for degradation, and *SLC30A1* encodes the sequence for a zinc transporter that inhibits calcium channel activity. The Search Tool for the Retrieval of Interacting Genes/ Proteins (STRING) analysis predicted *GPR180* to functionally interact with either the choroidermia-like gene *(CHML)* or TBC1 domain 8 *(TBC1D8)* gene to activate Rab family proteins or lysosomal-associated membrane protein 3 *(LAMP3)* gene, which further implicates the $1,25D_3$-MARRS receptor in $1,25D_3$-stimulated vesicular transport.

Investigators have shown that the $1,25D_3$-MARRS receptor has both nuclear and endoplasmic reticulum retention motifs [6] that can facilitate its translocation from the plasma membrane to the Golgi Apparatus [25], the endoplasmic reticulum where it is shown to be identical to the endoplasmic reticulum p57 (Erp57) [6] or the nucleus [16, 17, 23, 27] to elicit genomic cellular events, which occur through a series of vesicular transport events, with the involvement of cytoskeletal components such as microtubules, micro-filaments, and intermediate filaments.

Table 1. DNA identified from DNA-1,25D$_3$-MARRS complexes obtained using chromatin-immunoprecipitation experiments of chick enterocytes exposed to vehicle or hormone. DNA isolates were sequenced and identified using ExPASy Bioinformatics Resource Portal

DNA isolate (Protein name)	Accession #	Molecular function	Predicted functional partners (Genes)[*]
Vehicle (0.01% ethanol)			
HOMER-1 (Homer protein homologue-1)	NM 152134.2 NM 147176.2	Post-synaptic densi_y scaffolding protein; aids in coupling surface receptors to intracellular calcium release; regulates calcium channel activated receptors in skeletal and cardiac muscle	*APP, DLG4, GRIN3A, GRM1, GRM5, HOMER2, ITPR1, MIF, MYO18B, SHANK1*
PDIA3 (protein disulphide isomerase-A3)	NM 007952.2	Enzyme that catalyses protein folding	*CALR, CANX, HSPA5, IL2, IL2RA PADI2, RARA, SRC, STAT3, TAPBP*
Hormone (300 pM 1,25D$_3$)			
GPR180 (G protein-coupled receptor 180)	NM 021434.5	Integral membrane protein receptor	*CHML,GMDS, HARBI, LAMP3, POGZ, TBC1D8, UTRN*
NEURL1B (Neuralized homologue 1B)	NM001081656.2	E3 ubiquitin ligase	*DLK1,MIB1*
SLC30A1 (Zinc transporter 1)	NM 009579.3	Zinc ion transmembrane transporter activity; calcium channel inhibitor activity	*MT1, MTF1, SLC39A1, SLC39A2, SLC39A4, SLC39A7, SLC39A8, SLC39A10, SLC39A13*

[*]The predicted functional partners of the DNA isolates were identified using STRING (Search Tool for the Retrieval of Interacting Genes/Proteins) 9.0 Database.

While Nemere and Szego [4] have previously identified tubulin as a key cytoskeletal component in $1,25D_3$-mediated calcium uptake, we have recently identified both actin and keratin as proteins that specifically interact with the $1,25D_3$-MARRS receptor before and after exposure to hormone in chick enterocytes.

We performed co-immunoprecipitation (co-IP) studies and identified actin and keratin through mass spectroscopic analysis as proteins that associate with the $1,25D_3$-MARRS receptor. Co-IP of both actin and keratin was also confirmed by the Western blot analysis (data not shown).

Using confocal microscopy, we investigated $1,25D_3$-stimulated changes in intracellular redistribution of the $1,25D_3$-MARRS receptor and simultaneous involvement of cytoskeletal actin (F-actin) in chick enterocytes. Using immunofluorescent labelling techniques, we demonstrated that $1,25D_3$ induced an increase in intracellular $1,25D_3$-MARRS receptor within 5 min) with observed cyclical changes in F-actin (Sterling and Nemere, manuscript in preparation).

We also found $1,25D_3$ stimulated intracellular redistribution of keratin that paralleled intracellular redistribution of the $1,25D_3$-MARRS receptor(Sterling and Nemere, manuscript in preparation)). We can thus conclude from this evidence that both microfilaments and intermediate filaments are involved in intracellular, $1,25D_3$-stimulated redistribution of the $1,25D_3$-MARRS receptor.

Role of $1,25D_3$-MARRS Receptor on Select Disease States

Vitamin D is a multifunctional hormone and plays some critical roles in diseases and metabolic conditions. While many of its genomic roles associated to the classical VDR have been extensively investigated and will be the subject of future research, we have outlined some of the critical roles of vitamin D that are associated in relation to its membrane receptor, $ERp57/1,25D_3$-MARRS protein. This pivotal role is primarily associated with one of the fundamental processes in biological systems: the ability of proteins to fold into a pre-defined and functional conformation.

However, when such pre-defined and functional conformation deviates, the disease state arises, or *in utero* lethality occurs in the case of a homozygous deletion of $ERp57/1,25D_3$-MARRS receptor gene [45, 46]. Given that the $ERp57/1,25D_3$-MARRS receptor is critical for post-translational modification of proteins, in addition to its role as a molecular chaperone, and its requirement in the major histocompatability complex class I assembly and regulation of gene expression, it is highly probable that the protein is involved in some of the disease states, such as prion neurotoxicity, Alzheimer's disease, and cancer, among others. Moreover, its involvement in oxidative stress (reviewed in [47, 48]) makes it even more likely to be associated with such disease states. Here, we discuss some of the recent literatures related to the association of 1,25D$_3$-MARRS with select disease states that were not covered in a book chapter we contributed recently [47, 48].

In Cancer

Research has shown that 1,25D$_3$ inhibits growth of several cancer types, including breast cancer. It is primarily its classical receptor that has been implicated in the modulation of cancer by 1,25D$_3$, while the role of its membrane counterpart has not been investigated in detail. One recent study on Michigan Cancer Foundation-7 (MCF-7) breast cancer cells suggested 1,25D$_3$-MARRS receptor expression to interfere with the growth inhibitory activity of 1,25D$_3$, possibly through the nVDR [49]. In contrast, siRNA and genome-wide studies suggested that the anti-proliferative effects of 1,25D$_3$ in MCF-7 breast tumour cell lines do not rely on classical vitamin D pathways per se [50].

In a mouse model of skin carcinogenesis, long-term biological response generated by chronic dosing with a non-genomic–selective vitamin D steroid was observed [51]. They showed that both 1,25D$_3$ and 1α,25(OH)$_2$-lumisterol, a conformationally restricted analogue that can generate only non-genomic responses, are effective inhibitors of ultraviolet radiation (UVR) damage in an immunocompetent mouse (Skh:hr1) model susceptible to UV-induced tumours. Both 1,25D$_3$ and 1α,25(OH)$_2$-lumisterol significantly reduced UVR-induced cyclobutane pyrimidine dimer (it can be highly mutagenic if not repaired prior to cell division and can lead to UV-induced immunosuppression making them potentially carcinogenic), apoptotic sunburn cells and immunosuppression.

Furthermore, these compounds inhibited skin tumour development, both papillomas and squamous cell carcinomas, in mice *in vivo*.

They attributed these non-genomic physiological responses to VDR associated with plasma membrane caveolae. While we have refuted this theory and the reason why 1,25D$_3$-MARRS receptor is actually involved in this regard in some of our previous works [47, 52–55], a recent study with VDR and ERp57 in the photoprotective effects of 1,25D$_3$ against UVR-induced DNA damage in human skin fibroblasts, indicated that the VDR and ERp57 are directly associated outside the nucleus in which they mediate at least some of the non-genomic actions of 1,25D$_3$ [56].

In prostate cancer LNCaP cells, 1,25D$_3$ decreased invasiveness of cells by interaction with a putative membrane-associated receptor, which activated membrane-initiated signalling via the c-Jun N-terminal kinase (JNK)/stress-activated protein kinase (SAPK) MAPK signalling pathway [57]. Treatment with 1,25D$_3$ evoked a dose-dependent activation of the JNK/SAPK MAPK signalling pathways within 10 min, which demonstrated membrane-initiated signalling of 1,25D$_3$ in LNCaP cells. Furthermore, treatment with 1,25D$_3$ decreased LNCaP cell invasiveness by approximately 20% after 48 h. Using an inhibitor (SP600125) for the JNK/SAPK MAPK signalling pathway in combination with 1,25D$_3$ on LNCaP cells, the inhibitory action of 1,25D$_3$ on invasiveness was eliminated [56]. In a study, tissue cores from 164 gastric cancer patients were stained for ERp57 [58]. ERp57 staining was significantly decreased in cancer patients and metastases were compared with both normal gastric mucosa and metaplasias. The reduced ERp57 expression also correlated with greater depth of tumour invasion and advanced stage of the disease. Kaplan-Meier survival analysis determined that tumours with the highest quartile of ERp57 expression were statistically associated with longer post-operative survival, while a Cox proportional hazard analysis showed that maintenance of ERp57 expression was associated with longer post-operative survival. These results may provide useful prognostic information for gastric cancer patients. These results also suggested that the role of 1,25D$_3$ MARRS receptor in cancer patients may be modulated through multiple pathways and that it may be dependent on type, cell line or stage of metastasis. Further research should examine the potential for pharmacological or natural agents that modify 1,25D$_3$-MARRS expression or activity as anticancer agents.

In Alzheimer's Disease

Alzheimer's disease is one of the major diseases associated with ERp57 in which β-amyloids are the major components of the plaque observed in the brains of its patients, usually the elderly. The β-amyloids, though formed naturally, are cleared by the endoplasmic reticulum chaperone proteins, such as ERp57 during post-translational modification. Studies suggest that a chaperone declines, like ERp57, are one of the two central defects in

Alzheimer's disease [59], and deposits of β-amyloids are observed only in the elderly [60, 61]. Interestingly, the expression and activity of ERp57/1,25D3-MARRS receptor also declines in the intestine and kidney with age [5, 21, 22].

In a recent study, $1,25D_3$ MARRS receptor enhanced cerebral clearance of human β-amyloid peptide from mouse brain across the blood-brain barrier involving both genomic and non-genomic pathways [62]. Because forskolin enhanced β-amyloid elimination from the mouse brain [62], it is more likely to act through ERp57 than the classical VDR. In another study, the structure function results provided evidence that $1,25D_3$ activation of VDR-dependent genomic and non-genomic signalling work in combination to recover disregulated innate immune function in patients with Alzheimer's disease [63].

Recently, certain phytochemicals have been found to be associated with ERp57 function. One example is diosgenin, which is a plant-derived steroidal sapogenin. Mice treated with diosgenin significantly reduced amyloid plaques and neurofibrillary tangles in the cerebral cortex and hippocampus [64]. More importantly, $1,25D_3$-MARRS receptor was shown to be a target of diosgenin and $1,25D_3$-MARRS receptor knockdown completely inhibited diosgenin-induced axonal growth in cortical neurons. On the other hand, catechin (found widely in green tea, cinnamon and cocoa among others), particularly the galloylated one, is found to inhibit ERp57 binding to other macromolecular ligands [65]. This may be of greater importance in designing new polyphenol-based ERp57-specific inhibitors that may be useful in ameliorating disease states where ERp57 is known to aggravate the condition.

In addition, there are other reports where ERp57 is involved. One such finding is the requirement of ERp57 for platelet function, aggregation, haemostasis and thrombosis [66, 67]. Using enzyme activity function blocking antibodies, Holbrook et al. [67] demonstrated a role for ERp57 in platelet aggregation, dense granule secretion, fibrinogen binding, calcium mobilisation and thrombus formation under arterial conditions. A rabbit antibody generated against the protein strongly inhibited ERp57 in a functional assay and strongly inhibited platelet aggregation [65]. In another finding, ERp57 was suggested to play an early adaptive response in toxin-mediated stress-associated with Parkinson's disease [68], for which environmental factors are one among the causes. ERp57 is also implicated in heart failure. The protein is significantly upregulated in both animal and human models of right and left heart failure and appears to work through Src, a downstream target of c-Src kinase; and it was found that Src expression and phosphorylation were markedly upregulated in the failing ventricles [69]. For a brief discussion about the role of ERp57 on a few other diseases, such as diabetes, Wolcott Rallison syndrome and prion neurotoxicity, please refer to other recent reviews [47, 48].

Conclusion

$1,25D_3$ mediated rapid transport of calcium and phosphate ions in the intestine cells involving the 1,25D3-MARRS receptor. Given that the $1,25D_3$-MARRS protein has both cytoplasmic and nuclear motifs and is redistribution within the cell, evidence has emerged for its potential involvement through genomic action. Its involvement in disease states appears primarily through its chaperone action, post-translational modification of vital proteins and the ability to modulate oxidative stress.

Acknowledgments

This project was supported by the Utah Agricultural Experiment Station, Utah State University, and approved as journal paper number 8575.

References

[1]   Nemere, I., Garbi, N., Hämmerling, G. J., Khanal, R. C. Intestinal cell calcium uptake and the targeted knockout of the $1,25D_3$-MARRS (membrane-associated, rapid response steroid-binding) receptor/PDIA3/ Erp57. *J. Biol. Chem.* 2010 Oct;285(41):31859–66.

[2]   Nemere, I., Garcia-Garbi, N., Hämmerling, G. J., Winger, Q. Intestinal cell phosphate uptake and the targeted knockout of the $1,25D_3$-MARRS receptor/PDIA3/ERp57. *Endocrinology.* 2012 Apr;153(4):1609–15.

[3]   Meng, Y., Nemere, I. Effect of 24,25-dihydroxyvitamin D_3 on phosphate absorption in vivo. *Immunol. Endocrinol. Metabol. Agents Med. Chem.* 2013 Jan;13:60–67.

[4]   Nemere, I., Szego, C. M. Early actions of parathyroid hormone and 1,25-dihydroxycholecalciferol on isolated epithelial cells from rat intestine: I. Limited lysosomal enzyme release and calcium uptake. *Endocrinology.* 1981 Apr;108(4):1450–62.

[5]   Khanal, R. C., Peters, T. M., Smith, N. M., Nemere, I. Membrane receptor-initiated signaling in 1,25(OH)$_2$D$_3$-stimulated calcium uptake in intestinal epithelial cells. *J. Cell. Biochem.* 2008 Nov;105(4):1109–16.

[6]   Nemere, I., Farach-Carson, M. C., Rohe, B., Sterling, T. M., Norman, A. W., Boyan, B. D., et al. Ribozyme knockdown functionally links a 1,25(OH)2D3 membrane binding protein (1,25D3-MARRS) and phosphate uptake in intestinal cells. *Proc. Natl. Acad. Sci. US.* 2004a May; 101(19):7392–7.

[7]   Nemere, I. The 1,25D$_3$-MARRS protein: contribution to steroid-stimulated uptake in chicks and rats. *Steroids.* 2005 May–Jun;70(5–7): 455–7.

[8]   Nemere, I., Garbi, N., Hämmerling, G., Hintze, K. J. Role of the 1,25D$_3$-MARRS receptor in the 1,25(OH)$_2$D$_3$-stimulated uptake of calcium and phosphate in intestinal cells. *Steroids.* 2012 Aug;77(10):897–902.

[9]   Norman, A. W., Putkey, J. A., Nemere, I. Intestinal calcium transport: pleiotropic effects mediated by vitamin D. *Fed. Proc.* 1982 Jan;41 (1): 78–83.

[10]   Matsumoto, T., Fontaine, O., Rasmussen, H. Effect of 1,25-dihydroxyvitamin D3 on phospholipid metabolism in chick duodenal mucosal cell. Relationship to its mechanism of action. *J. Biol. Chem.* 1981 Apr;256(7):3354–60.

[11]   Brasitus, T. A., Dudeja, P. K., Eby, B., Lau, K. Correction by 1-25-dihydroxycholecalciferol of the abnormal fluidity and lipid composition of enterocyte brush border membranes in vitamin D-deprived rats. *J. Biol. Chem.* 1986 Dec;261(35):16404–9.

[12]   Mezzetti, G., Monti, M. G., Casolo, L. P., Piccinini, G., Moruzzi, M. S. 1,25-Dihydroxycholecalciferol-dependent calcium uptake by mouse mammary gland in culture. *Endocrinology.* 1988 Feb;122(2):389–94.

[13]   Nemere, I., Norman, A. W. Rapid action of 1,25-dihydroxyvitamin D$_3$ on calcium transport in perfused chick duodenum: effect of inhibitors. *J. Bone Miner. Res.* 1987 Apr;2(2):99–107.

[14]   Lieberherr, M., Grosse, B., Duchambon, P., Drüeke, T. A functional cell surface type receptor is required for the early action of 1,25-dihydroxyvitamin D$_3$ on the phosphoinositide metabolism in rat enterocytes. *J. Biol. Chem.* 1989 Dec;264(34):20403–6.

[15]   Bourdeau, A., Atmani, F., Grosse, B., Lieberherr, M. Rapid effects of 1,25-dihydroxyvitamin D$_3$ and extracellular Ca^{2+} on phospholipid metabolism in dispersed porcine parathyroid cells. *Endocrinology.* 1990 Dec; 127(6):2738–43.

[16] Nemere, I., Dormanen, M. C., Hammond, M. W., Okamura, W. H., Norman, A. W. Identification of a specific binding protein for 1 alpha,25-dihydroxyvitamin D_3 in basal-lateral membranes of chick intestinal epithelium and relationship to transcaltachia. *J. Biol. Chem.* 1994 Sep;269(38):23750–6.

[17] Nemere, I. Nongenomic effects of 1,25-dihydroxyvitamin D_3: potential relation of a plasmalemmal receptor to the acute enhancement of intestinal calcium transport in chick. *J. Nutr.* 1995 Jun;125(6 Suppl.): 1695S–8S.

[18] Nemere, I., Safford, S. E., Rohe, B., DeSouza, M. M., Farach-Carson, M. C. Identification and characterization of $1,25D_3$-membrane-associated rapid response, steroid ($1,25D_3$-MARRS) binding protein. *J. Steroid Biochem. Mol. Biol.* 2004b May;89–90(1–5):281–5.

[19] Larsson, D., Nemere, I., Sundell, K. Putative basal lateral membrane receptors for 24,25-dihydroxyvitamin D(3) in carp and Atlantic cod enterocytes: characterization of binding and effects on intracellular calcium regulation. *J. Cell. Biochem.* 2001 Aug;83(2):171–86.

[20] Larsson, D., Nemere, I., Aksnes, L., Sundell, K. Environmental salinity regulates receptor expression, cellular effects, and circulating levels of two antagonizing hormones, 1,25-dihydroxyvitamin D_3 and 24,25-dihydroxyvitamin D_3, in rainbow trout. *Endocrinology.* 2003 Feb;144 (2): 559–66.

[21] Larsson, B., Nemere, I. Effect of growth and maturation on membrane-initiated actions of 1,25-dihydroxyvitamin D(3). I. Calcium transport, receptor kinetics, and signal transduction in intestine of male chickens. *Endocrinology.* 2003a May;144(5):1726–35.

[22] Larsson, B., Nemere, I. Effect of growth and maturation on membrane-initiated actions of 1,25-dihydroxyvitamin D_3-II: calcium transport, receptor kinetics, and signal transduction in intestine of female chickens. *J. Cell. Biochem.* 2003b Dec;90(5):901–13.

[23] Rohe, B., Safford, S. E., Nemere, I., Farach-Carson, M. C. Identification and characterization of $1,25D_3$-membrane-associated rapid response, steroid ($1,25D_3$-MARRS)-binding protein in rat IEC-6 cells. *Steroids.* 2005 May–Jun;70(5–7):458–63.

[24] Mesbah, M., Nemere, I., Papagerakis, P., Nefussi, J. R., Orestes-Cardoso, S., Nessmann, C., et al. Expression of a 1,25-dihydroxyvitamin D_3 membrane-associated rapid-response steroid binding protein during human tooth and bone development and biomineralization. *J. Bone Miner. Res.* 2002 Sep;17(9):1588–96.

[25] Jia, Z., Nemere, I. Immunochemical studies on the putative plasmalemmal receptor for 1,25-dihydroxyvitamin D_3 II. Chick kidney and brain. *Steroids*. 1999 Aug;64(8):541–50.

[26] Boyan, B. D., Sylvia, V. L., Dean, D. D., Pedrozo, H., Del Toro, F., Nemere, I., et al. 1,25-$(OH)_2D_3$ modulates growth plate chondrocytes via membrane receptor-mediated protein kinase C by a mechanism that involves changes in phospholipid metabolism and the action of arachidonic acid and PGE2. *Steroids*. 1999 Jan–Feb;64(1–2):129–36.

[27] Nemere, I., Ray, R., McManus, W. Immunochemical studies on the putative plasmalemmal receptor for 1, 25(OH)(2)D(3). I. Chick intestine. *Am. J. Physiol. Endocrinol. Metab.* 2000 Jun;278(6):E1104–14.

[28] Nemere, I., Campbell, K. Immunochemical studies on the putative plasmalemmal receptor for 1, 25-dihydroxyvitamin D(3). III. Vitamin D status. *Steroids*. 2000 Aug;65(8):451–7.

[29] Berdal, A., Mesbah, M., Papagerakis, P., Nemere, I. Putative membrane receptor for 1,25$(OH)_2$ vitamin D_3 in human mineralized tissues during prenatal development. *Connect Tissue Res*. 2003;44(Suppl. 1):136–40.

[30] Teillaud, C., Nemere, I., Boukhobza, F., Mathiot, C., Conan, N., Oboeuf, M., et al. Modulation of 1alpha,25-dihydroxyvitamin D_3-membrane associated, rapid response steroid binding protein expression in mouse odontoblasts by 1alpha,25-$(OH)_2D_3$. *J. Cell. Biochem*. 2005 Jan;94(1): 139–52.

[31] Sutherland, S. K., Nemere, I., Benishin, C. G. Regulation of parathyroid hypertensive factor secretion by vitamin D_3 analogs in parathyroid cells derived from spontaneously hypertensive rats. *J. Cell. Biochem*. 2005 Sep; 96(1):97–108.

[32] Khanal, R. C., Smith, N. M., Nemere, I. Phosphate uptake in chick kidney cells: effects of 1,25$(OH)_2D_3$ and 24,25$(OH)_2D_3$. *Steroids*. 2007 Feb; 72(2):158–64.

[33] Schwartz, Z., Sylvia, V. L., Larsson, D., Nemere, I., Casasola, D., Dean, D. D., et al. 1alpha,25$(OH)_2D_3$ regulates chondrocyte matrix vesicle protein kinase C (PKC) directly via G-protein-dependent mechanisms and indirectly via incorporation of PKC during matrix vesicle biogenesis. *J. Biol. Chem*. 2002 Apr;277(14):11828–37.

[34] Nemere, I., Schwartz, Z., Pedrozo, H., Sylvia, V. L., Dean, D. D., Boyan, B. D. Identification of a membrane receptor for 1,25-dihydroxyvitamin D_3 which mediates rapid activation of protein kinase C. *J. Bone Miner. Res*. 1998 Sep;13(9):1353–9.

[35] Boyan, B. D., Bonewald, L. F., Sylvia, V. L., Nemere, I., Larsson, D., Norman, A. W., et al. Evidence for distinct membrane receptors for 1 alpha,25-(OH)(2)D(3) and 24R,25-(OH)(2)D(3) in osteoblasts. *Steroids.* 2002 Mar;67(3–4):235–46.

[36] Zhao, B., Nemere, I. 1,25(OH)$_2$D$_3$-mediated phosphate uptake in isolated chick intestinal cells: effect of 24,25(OH)$_2$D$_3$, signal transduction activators, and age. *J. Cell. Biochem.* 2002;86(3):497–508.

[37] Sterling, T. M., Nemere, I. 1,25-dihydroxyvitamin D$_3$ stimulates vesicular transport within 5 s in polarized intestinal epithelial cells. *J. Endocrinol.* 2005 Apr;185(1):81–91.

[38] De Boland, A. R., Norman, A. W. 1alpha,25(OH)$_2$-vitamin D$_3$ signaling in chick enterocytes: enhancement of tyrosine phosphorylation and rapid stimulation of mitogen-activated protein (MAP) kinase. *J. Cell. Biochem.* 1998 Jun; 69(4):470–82.

[39] Schwartz, Z., Ehland, H., Sylvia, V. L., Larsson, D., Hardin, R. R., Bingham, V., et al. 1alpha,25-dihydroxyvitamin D(3) and 24R,25-dihydroxyvitamin D(3) modulate growth plate chondrocyte physiology via protein kinase C-dependent phosphorylation of extracellular signal-regulated kinase 1/2 mitogen-activated protein kinase. *Endocrinology.* 2002 Jul;143(7):2775–86.

[40] Pardo, V. G., Boland, R., de Boland, A. R. 1alpha,25(OH)(2)-Vitamin D(3) stimulates intestinal cell p38 MAPK activity and increases c-Fos expression. *Int. J. Biochem. Cell. Biol.* 2006;38(7):1181–90.

[41] Pardo, V. G., de Boland, A. R. Tyrosine phosphorylation signalling dependent on 1alpha,25(OH)$_2$-vitamin D$_3$ in rat intestinal cells: effect of ageing. *Int. J. Biochem. Cell. Biol.* 2004 Mar;36(3):489–504.

[42] Rohe, B., Safford, S. E., Nemere, I., Farach-Carson, M. C. Regulation of expression of 1,25D$_3$-MARRS/ERp57/PDIA3 in rat IEC-6 cells by TGF beta and 1,25(OH)$_2$D$_3$. *Steroids.* 2007 Feb;72(2):144–50.

[43] Wu, W., Beilhartz, G., Roy, Y., Richard, C. L., Curtin, M., Brown, L., et al. Nuclear translocation of the 1,25D$_3$-MARRS (membrane associated rapid response to steroids) receptor protein and NFkappaB in differentiating NB4 leukemia cells. *Exp. Cell. Res.* 2010 Apr;16(7): 1101–8.

[44] Grindel, B. J., Rohe, B., Safford, S. E., Bennett, J. J., Farach-Carson, M. C. Tumor necrosis factor-α treatment of HepG2 cells mobilizes a cytoplasmic pool of ERp57/1,25D$_3$-MARRS to the nucleus. *J. Cell. Biochem.* 2011 Sep;112(9):2606–15.

[45] Garbi, N., Tanaka, S., Momburg, F., Hämmerling, G. J. Impaired assembly of the major histocompatibility complex class I peptide-loading complex in mice deficient in the oxidoreductase ERp57. *Nat. Immunol.* 2006 Jan;7(1):93–102.

[46] Wang, Y., Chen, J., Lee, C. S., Nizkorodov, A., Riemenschneider, K., Martin, D., et al. Disruption of Pdia3 gene results in bone abnormality and affects 1α,25-dihydroxy-vitamin D₃-induced rapid activation of PKC. *J. Steroid Biochem. Mol. Biol.* 2010 Jul;121(1–2):257–60.

[47] Khanal, R. C., Nemere, I. Membrane receptors for vitamin D metabolites and the role of reactive oxygen species. In: Gombart A, editor. *Vitamin D: Oxidation, Immunity and Aging.* London, UK: CRC Press, Taylor and Francis Group; 2012.p201–19.

[48] Sterling, T., Khanal, R. C., Meng, Y., Zhang, Y., Nemere, I. Functional consequences of vitamin D metabolites and their receptors. In: Dakshinamurti, K., Dakshinmurti, S., editors *Vitamin-Binding Proteins - Their Functional Consequences.* London, UK: CRC Press, Taylor and Francis Group;2013. In press.

[49] Richard, C. L., Farach-Carson, M. C., Rohe, B., Nemere, I., Meckling, K. A. Involvement of 1,25D₃-MARRS (membrane associated, rapid response steroid-binding), a novel vitamin D receptor, in growth inhibition of breast cancer cells. *Exp. Cell. Res.* 2010 Mar;316(5):695–703.

[50] Costa, J. L., Eijk, P. P., van de Wiel, M. A., ten Berge, D., Schmitt, F., Narvaez, C. J., et al. Anti-proliferative action of vitamin D in MCF7 is still active after siRNA-VDR knock-down. *BMC Genomics.* 2009 Oct; 10: 499.

[51] Dixon, K. M., Norman, A. W., Sequeira, V. B., Mohan, R., Rybchyn, M. S., Reeve, V. E., et al. 1α,25(OH)₂-vitamin D and a nongenomic vitamin D analogue inhibit ultraviolet radiation-induced skin carcinogenesis. *Cancer Prev. Res.* (Phila). 2011 Sep;4(9):1485–94.

[52] Khanal, R., Nemere, I. Membrane receptors for vitamin D metabolites. *Crit. Rev. Eukaryot. Gene Expr.* 2007;17(1):31–47.

[53] Khanal, R. C., Nemere, I. The ERp57/GRp58/1,25D₃-MARRS receptor: multiple functional roles in diverse cell systems. *Curr. Med. Chem.* 2007; 14(10):1087–93.

[54] Khanal, R. C., Nemere, I. Regulation of intestinal calcium transport. *Annu. Rev. Nutr.* 2008;28:179–96.

[55] Khanal, R. C., Nemere, I. Newly recognized receptors for vitamin D metabolites. *Immunol. Endocrine Metabol. Agents Med. Chem.* 2009;9: 143–52.

[56] Sequeira, V. B., Rybchyn, M. S., Tongkao-On, W., Gordon-Thomson, C., Malloy, P. J., Nemere, I., et al. The role of the vitamin D receptor and ERp57 in photoprotection by 1α,25-dihydroxyvitamin D$_3$. *Mol. Endocrinol.* 2012 Apr;26(4):574–82.

[57] Larsson, D., Hagberg, M., Malek, N., Kjellberg, C., Senneberg, E., Tahmasebifar, N., et al. Membrane initiated signaling by 1,25alpha-dihydroxyvitamin D$_3$ in LNCaP prostate cancer cells. *Adv. Exp. Med. Biol.* 2008;617:573–9.

[58] Leys, C. M., Nomura, S., LaFleur, B. J., Ferrone, S., Kaminishi, M., Montgomery, E., et al. Expression and prognostic significance of prothymosin-alpha and ERp57 in human gastric cancer. *Surgery.* 2007 Jan;141(1):41–50.

[59] Erickson, R. R., Dunning, L. M., Holtzman, J. L. The effect of aging on the chaperone concentrations in the hepatic, endoplasmic reticulum of male rats: the possible role of protein misfolding due to the loss of chaperones in the decline in physiological function seen with age. *J. Gerontol. A Biol. Sci. Med. Sci.* 2006 May;61(5):435–43.

[60] Dobson, C. M. Protein aggregation and its consequences for human disease. *Protein Pept. Lett.* 2006;13(3):219–27.

[61] Thomas, P. J., Qu, B. H., Pedersen, P. L. Defective protein folding as a basis of human disease. *Trends Biochem. Sci.* 1995 Nov;20(11):456–9.

[62] Ito, S., Ohtsuki, S., Nezu, Y., Koitabashi, Y., Murata, S., Terasaki, T. 1α,25-Dihydroxyvitamin D$_3$ enhances cerebral clearance of human amyloid-β peptide(1-40) from mouse brain across the blood-brain barrier. *Fluids Barriers CNS.* 2011 Jul;8:20.

[63] Mizwicki, M. T., Menegaz, D., Zhang, J., Barrientos-Durán, A., Tse, S., Cashman, J. R., et al. Genomic and nongenomic signaling induced by 1α,25(OH)$_2$-vitamin D$_3$ promotes the recovery of amyloid-β phagocytosis by Alzheimer's disease macrophages. *J. Alzheimers Dis.* 2012; 29(1):51–62.

[64] Tohda, C., Urano, T., Umezaki, M., Nemere, I., Kuboyama, T. Diosgenin is an exogenous activator of 1,25D$_3$-MARRS/Pdia3/ERp57 and improves Alzheimer's disease pathologies in 5XFAD mice. *Sci. Rep.* 2012;2:535.

[65] Trnková, L., Ricci, D., Grillo, C., Colotti, G., Altieri, F. Green tea catechins can bind and modify ERp57/PDIA3 activity. *Biochim. Biophys. Acta* 2012 Nov.

[66] Wu, Y., Ahmad, S. S., Zhou, J., Wang, L., Cully, M. P., Essex, D. W. The disulfide isomerase ERp57 mediates platelet aggregation, hemostasis, and thrombosis. *Blood.* 2012 Feb;119(7):1737–46.

[67] Holbrook, L. M., Sasikumar, P., Stanley, R. G., Simmonds, A. D., Bicknell, A. B., Gibbins, J. M. The platelet-surface thiol isomerase enzyme ERp57 modulates platelet function. *J. Thromb. Haemost.* 2012 Feb;10(2):278–88.

[68] Kim-Han, J. S., O'Malley, K. L. Cell stress induced by the parkinsonian mimetic, 6-hydroxydopamine, is concurrent with oxidation of the chaperone, ERp57, and aggresome formation. *Antioxid. Redox. Signal.* 2007 Dec;9(12):2255–64.

[69] Vitello, A. M., Du, Y., Buttrick, P. M., Walker, L. A. Serendipitous discovery of a novel protein signaling mechanism in heart failure. *Biochem. Biophys. Res. Commun.* 2012 May;421(3):431–5.

In: New Developments in Calcium ...   ISBN: 978-1-62948-601-7
Editor: Masayoshi Yamaguchi   © 2014 Nova Science Publishers, Inc.

Chapter II

Modulation of Airway Epithelial Cell Calcium and Airway Hydration in Cystic Fibrosis: Role of Glucocorticoids and Lipoxins

*Fiona C. Ringholz[1], Gerard Higgins[1],
Julien M. Buyck[2] and Valérie Urbach[1,3,4]**

[1]National Children's Research Centre, Crumlin, Dublin, Ireland
[2]Pharmacologie Cellulaire et Moléculaire, Université
Catholique de Louvain, Bruxelles, Belgium
[3]Institut National de la Santé et de la Recherche Médicale, Paris
[4]Faculté de Médecine, Université Paris Descartes, Paris, France

Abstract

Cystic Fibrosis (CF) is characterised by ion transport abnormalities caused by a mutation of the Cystic Fibrosis Transmembrane conductance

* Corresponding author: Institut National de la Santé et de la Recherche Médicale, U845, Faculté de Médecine, Université Paris Descartes,156 rue Vaugirard 75015, Paris, France. E-mail: valerie.urbach@ncrc.ie; Telephone:+ 353 (01) 4096419 or 6585; Fax: + 353 (01) 4550201.

Regulator (CFTR) gene. This results in airway surface dehydration, impaired muco-ciliary clearance, favours chronic bacterial infection and sustained inflammation and leads to progressive lung destruction.

In CF, intracellular calcium status represents a double-edged sword. Indeed, increases in intracellular calcium can restore the airway surface liquid layer architecture and muco-ciliary clearance via stimulation of the calcium-activated chloride currents [1, 2]. However, heightened intracellular calcium activity has been implicated as a major participant in promoting hyper-inflammatory pathways via its role in endoplasmic reticulum biology and its second messenger function, interacting with calcium-sensitive proteins and promoting nuclear transcription of pro-inflammatory mediators [3]. In this chapter, we explore how glucocorticoids and lipoxins differentially regulate intracellular calcium concentration, airway hydration and inflammation in CF lung disease. In particular, we highlight how Lipoxin A_4 (LXA$_4$) couples the beneficial effects of raised intracellular calcium concentration to promote chloride secretion and airway hydration [2] to an anti-inflammatory and pro-resolution programme of transcription [4].

Introduction

CF is the most common lethal inherited disease in Caucasians, characterised by a defect in the CFTR which encodes a chloride channel. CF affects many organs but the progressive lung destruction is the main cause of morbidity and mortality. Loss of CFTR function leads to defective chloride (Cl$^-$) secretion and sodium (Na$^+$) hyperabsorption resulting in a reduced Airway Surface Liquid (ASL) height.

This impairs muco-ciliary clearance and innate immune defence mechanisms in the airways, favouring chronic bacterial colonization. A cycle of persistent infection and hyper inflammation is set up, eventually manifesting as bronchiectasis and progressive lung destruction [5].

A central difficulty in attempting to ameliorate the basic defect of CFTR dysfunction in CF pertains to regulation of intracellular calcium concentration. In the absence of functional CFTR the main compensatory strategy to promote ASL hydration relies on stimulating alternative chloride secretory pathways such as the activity of Calcium Activated Chloride Channels (CaCC) by raised intracellular concentrations of calcium. However, the hyper-inflammatory phenotype of the airway epithelial cell in CF has been mechanistically attributed to the expansion of intracellular calcium stores and amplification of the calcium dependant inflammatory response [3].

Normal Calcium Signalling in Airway Epithelia

Calcium is a ubiquitous intracellular messenger able to regulate a wide range of essential cellular functions of respiratory epithelium including, cytoskeleton and tight junction organizations, ciliary beat frequency [6-8], secretion of mucus and ion transport [1, 9].

Normal calcium signalling in airway epithelia is illustrated in Figure 1A. The intracellular calcium concentration ($[Ca]_i$) is tightly regulated and remains relatively low (100 nM) compared to the extracellular concentration (2 mM). A diverse array of extracellular stimuli communicate with the airway epithelial cell via G-Protein Coupled Receptors (GPCR) stimulating an influx of Ca^{2+} into the cell from the extracellular space.

GPCRs form multimeric protein complexes including with Phospholipase C (PLC) [10] which serves to cleave the lipid PIP_2 in two separate second messengers, Diacylglycerol (DAG) and Inositol triphosphate (IP_3), each acting on different targets. DAG activates Protein Kinase C (PKC) and some Transient Receptor Potential Canonical (TRPC) channels involved in extracellular calcium influx. IP_3 activates specific receptors (IP_3R) and stimulates Ca^{2+} release from endoplasmic reticulum (ER).

The cytosolic change in Ca^{2+} is biphasic, with an initial transient rise followed by a sustained elevation [11].

The initial transient phase is induced by IP_3-stimulated release of Ca^{2+} from intracellular stores and is transient due to the removal of Ca^{2+} by subsequent activation of the plasma membrane Ca^{2+} pump ATPase (PMCA).

Subsequently, Ca^{2+} release from intracellular stores activates sustained Ca^{2+} influx through channels in the plasma membrane, known as store-operated Ca^{2+} entry (SOCE).

Indeed, depletion of Ca^{2+} from the endoplasmic reticulum (ER) triggers a signalling cascade by activating the ER resident Ca^{2+} sensor protein stromal interacting molecule 1 (STIM1) [12-15].

STIM 1 activates the Ca^{2+} release activated channels (CRAC) at the plasma membrane, Orai1 [16, 17] and TRPC6, endogenously expressed in human airway epithelial cells [18, 19]. TRPC6 is known to form selective and non-selective divalent cation channels activated by Diacylglycerols (DAG) [18, 20]. Removal of $[Ca]_i$ from the cytoplasm depends on the activation of the SERCA2 pump [21] to replenish ER and also upon PMCA, to extrude calcium towards the extracellular space [22].

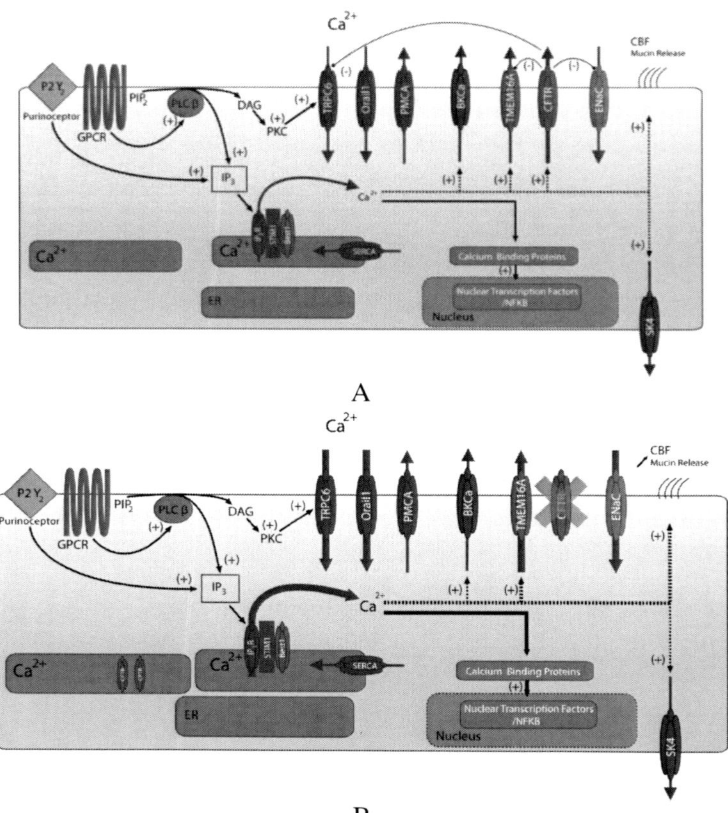

A

B

Figure 1. (A) Calcium signaling and regulation in normal airway epithelial cells. (B) Calcium signaling and regulation in CF airway epithelial cells. (A) P2Y$_2$ and GPCR stimulation generate increased intracellular calcium concentration. GPCR stimulation activates PLC generating DAG and IP$_3$. DAG activates PKC, which in turn activates TRPC6, and extracellular calcium influx. IP$_3$ activates IP$_3$Rs stimulating calcium release from the ER. Depletion of calcium from the ER activates STIM1which, in turn activates Orai1 and TRPC6 and extracellular calcium influx. Calcium is removed from the cytosol by PMCA extruding calcium towards the extracellular space and by the SERCA pump replenishing ER stores. Raised intracellular calcium concentration affects cellular potassium transport (Purple), chloride secretion (Blue), sodium absorption (Green) and calcium sensitive nuclear signalling cascades, particularly NFκB activation. (B) GPCR induced calcium mobilisation is amplified in CF airway epithelial cells. Calcium release from intracellular stores and entry through the plasma membrane are increased. ER calcium stores are expanded. Excessive NFκB activation in CF has been attributed to abnormal intracellular calcium regulation and amplification of calcium sensitive pro-inflammatory signalling cascades. Abreviations: P2Y$_2$, P2Y Purinoceptor 2; GPCR, G-Protein Coupled Receptor; PIP$_2$, Phosphatidylinositol 4,5-bisphosphate; PLCß, Phospholipase Cß; IP$_3$, Inositol Trisphosphate; DAG, Diacylglycerol; PKC, Protein inase C; ER, Endoplasmic Reticulum; TRPC6, Transient Receptor Potential Cation Channel, C6; Orai1, Calcium release-activated calcium channel protein 1; PMCA, Plasma Membrane Calcium ATPase; BKCa, Ligand-sensitive Potassium Large Conductance Calcium-activated Ion Channel; TMEM16A, Anoctamin-1 Calcium Activated Chloride Channel; CFTR, Cystic Fibrosis Transmembrane Conductance Regulator; ENaC, Epithelial Sodium Channel; CBF, Ciliary Beat Frequency; IP$_3$R, Inositol Trisphosphate Receptor; STIM1, Stromal Interaction Molecule 1; Best1, Bestrophin-1; SERCA, Sarco/Endoplasmic Reticulum Ca^{2+}-ATPase; SK4, Potassium intermediate/small Conductance Calcium-activated Channel.

Abnormal Calcium Signalling in Cystic Fibrosis

Several aspects of Ca^{2+} signalling are disturbed in CF airway epithelial cells when compared to non-CF human airway epithelial cells, illustrated in Figure 1B. GPCR induced calcium mobilisation is amplified in CF airway epithelial cells, more specifically, Ca^{2+} release from intracellular stores and entry through the plasma membrane are increased [23]. IP₃Rs-dependent Ca^{2+} release is abnormally increased in CF epithelial cells [24], as a consequence of misfolded mutated CFTR protein trapped in the ER [25]. ER calcium store expansion has been described associated with ER concentration and condensation around the nucleus [23, 25]. This results in IP₃R clustering at the ER membrane, facilitating their activation. In support of these findings, miglustat treatment of CF airway epithelial cells restores F508del-CFTR to the apical plasma membrane, and is associated with normalisation of Ca^{2+} homeostasis [25].

Another hypothesis suggests that differences in calcium signalling in CF cells occur independently of the CFTR defect, following the observation that chronic exposure of non-CF airway epithelial cells to bacterial stimulation results in increased ER size and Ca^{2+} storage [23].

In agreement with this hypothesis, it has recently been confirmed that *Pseudomonas aeruginosa* is able to induce an intracellular Ca^{2+} increase in airway epithelial cells [26, 27].

A member of the TRPC channel family, TRPC6, involved in extracellular calcium influx at the plasma membrane, has been attributed with responsibility for the abnormal increase of Ca^{2+} influx in human airway epithelial CF cells. TRPC6 and CFTR channels are both present within a multi-protein complex in which each channel regulates the other [18]. In the same manner that has been described for epithelial Na^+ channel (ENaC), CFTR exerts a regulatory effect upon TRPC6, supressing TRPC6-dependent Ca^{2+} influx [28]. In CF cells, the loss of functional CFTR expression at the plasma membrane removes this interaction and increases TRPC6 dependent Ca^{2+} influx. Moreover, enhanced Orai1/STIM1 complex formation during store depletion in CF cells also contributes to increased Ca^{2+} signalling [29].

Finally, 14-3-3 proteins were reported to be up regulated in CF epithelial cells [30]. 14-3-3 proteins inhibit the activity of the PMCA, increasing $[Ca]_i$, and are known to inhibit CaCC (via a calmodulin modulated pathway) thus inhibiting Cl⁻ secretion [31].

Calcium in Airway Epithelial Inflammation

Calcium plays a critical role in cell physiology, regulating diverse cellular functions via its second messenger function. Intracellular calcium concentration affects almost all known signalling pathways via its interaction with calcium sensitive enzymes and proteins. Examples of such proteins participating in signalling cascades include kinases, phosphatases, transcription factors, calcineurin and calmodulin.

The immune system senses pathogens via conserved molecular patterns interacting with Pattern Recognition Receptors (PRR's) such as the Toll-like Receptor family [32]. These receptors are coupled to signal transduction pathways that control inducible immune response genes. An example of such a signalling pathway in airway epithelial cells comes from the report that TLR5 in co-operation with either TLR2 or asialo-GM1 receptor binds bacterial flagellum from *Pseudomonas aeruginosa* to transduce a MyD88-dependent signalling cascade that activates protein kinases and a calcium signalling pathway [33]. This cascade results in IL8 induction via phosphorylation of MAPK's p38, ERK and JNK and activation of transcription factors, in particular NFκB [33].

The NFκB family of transcription factors are powerful regulators of inducible gene expression coordinating both innate and adaptive responses to antigenic stimuli, resulting in cytokine responses and ultimately in pathogen clearance. IL8 is a potent chemotactic peptide that induces the expression of adhesion molecules and recruits and activates neutrophils at sites of inflammation [34, 35]. IL8 expression is amplified in CF airways and negatively correlated with lung function [36] and its transcription is controlled by several nuclear transcription factors, namely CREB, CHOP, AP-1, NF-IL6 and NFκB [33].

Excessive NFκB activation has been observed in CF and variously attributed to; ER Ca^{2+} store expansion and amplification of the Ca^{2+} dependent inflammatory response; ER stress due to the accumulation of misfolded CFTR protein; persistent bacterial infection and pro-inflammatory cytokine stimulation; and intrinsic imbalances between pro and anti-inflammatory cytokine profiles in CF [37]. In CF, enhanced activation of NFκB permits translocation of NFκB to the nucleus where it binds to its site in the promoter region of the IL-8 gene and rapidly up-regulates IL8 synthesis.

Anti-Inflammatory Role
for Glucocorticoids

Glucocorticoids are a class of steroid hormone synthesised in the adrenal cortex and released into circulation in response to a wide range of stressful stimuli. Their release is co-ordinated by the Hypothalamic-Pituitary-Adrenal axis and they possess a vast array of functions as exemplified by the fact that an organism cannot survive without them. They are potent suppressors of inflammation, and as such, synthetic glucocorticoids are central to anti-inflammatory therapy in several inflammatory disorders [38].

Glucocorticoids bind to their cognate intracellular nuclear hormone receptors. They exert direct effects on gene expression by the binding of glucocorticoid receptors (GR) to glucocorticoid-responsive elements and indirect effects through protein interference interactions of glucocorticoid receptors with transcription factors such as NFκB and activator protein 1 [38], illustrated in Figure 2A. For example, dexamethasone (a synthetic glucocorticoid) antagonises IL8 transcription by binding to and activating the GR. The activated GR binds directly to NFκB, bound to its own site in the IL8 gene promoter. This interaction physically hinders the phosphorylation of RNA polymerase II − a step which is necessary for the initiation of transcription [39].

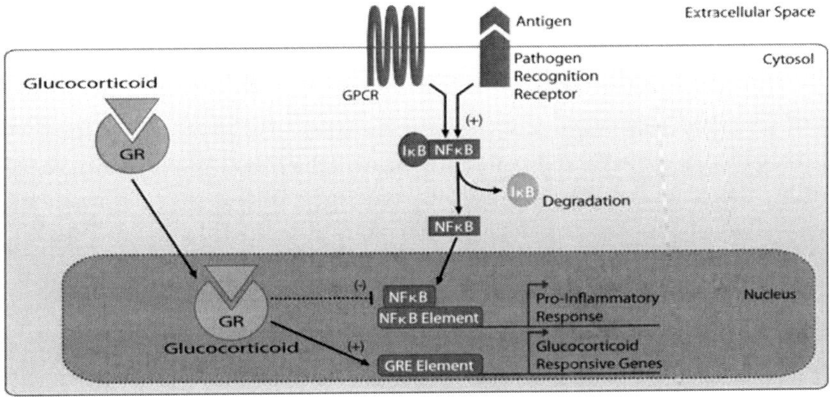

Figure 2. (Continued).

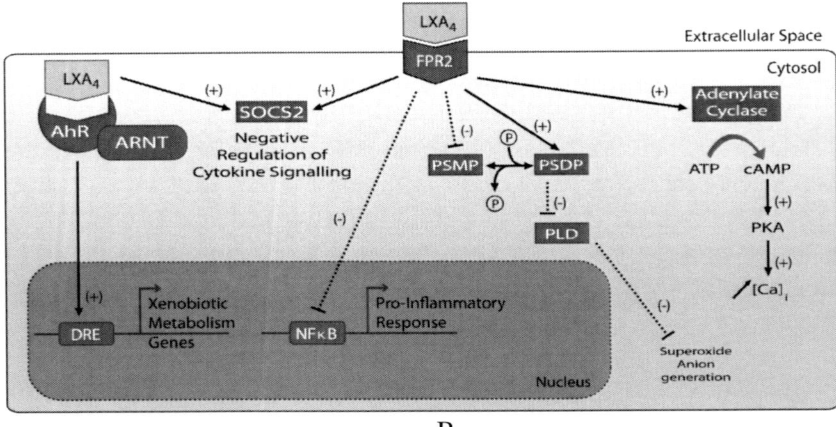

B

Figure 2. Anti-inflammatory signalling of (A) Glucocorticoids and (B) Lipoxin A$_4$. (A) Antigens and pro-inflammatory immune signals stimulate IκB degradation releasing NFκB which translocates to its binding site in the nucleus. Glucocorticoids ligate the GR which binds to the GRE Element and activates transcription of Glucocorticoid Responsive genes. The ligated GR indirectly affects NFκB pro-inflammatory transcription via protein interference interactions. (B) LXA$_4$ interacts with FPR2 and AhR receptors. AhR dimerizes with ARNT and regulates genes involved in xenobiotic metabolism. LXA$_4$ stimulates [Ca]$_i$ increase via adenylate cyclase activity and PKA activation. LXA$_4$ reduces transcription of genes under NFκB control. LXA$_4$ stimulation leads to PSDP accumulation in favour of PSMP, inhibiting PLD activation and superoxide anion generation. LXA$_4$ stimulation results in AhR and FPR2 dependent Socs2 upregulation and thus negative regulation of cytokine signalling. Abreviations: GR, Glucocorticoid Receptor; GPCR, G-Protein Coupled Receptor; GRE, Glucocorticoid-Responsive Element; LXA$_4$, Lipoxin A$_4$; AhR, Aryl hydrocarbon receptor; ARNT, AhR Nuclear Transporter; DRE, Dioxin Response Elements; PSDP, Presqualene Diphosphate; PSMP, Presqualene Monophosphate; PLD, Phospholipase D; PKA, Protein Kinase A; Socs2, Suppressor of Cytokine Signalling 2.

In CF, the use of glucocorticoids has been limited by adverse side effect profiles when taken systemically [40] (cataracts and growth retardation) and disappointingly equivocal efficacy when administered via inhalation [41].

Anti-Inflammatory Role for Lipoxin A$_4$

Lipoxins are endogenously produced lipid mediators which display diverse and potent anti-inflammatory actions, limiting pro-inflammatory responses and promoting the return of inflamed tissue to homeostasis [4]. Lipoxins are the first of the Specialised Pro-Resolution Mediators to be produced as inflammatory exudates advance from propagation of acute inflammation to its active resolution [42]. They act as stop signals for inflammation mediating their effects on both recruited immune effector cells and resident cells. In particular, Lipoxin A_4 (LXA_4) stops neutrophil recruitment [43], stimulates monocyte recruitment [44] and enhances neutrophil clearance by macrophages [45] (extensively reviewed in [46]). Mice treated with analogues of LXA_4 and subsequently challenged with *P. aeruginosa* contained that bacterial challenge more effectively [47]. In human differentiated bronchial epithelial cells, LXA_4 antagonises TNFα induced IL8 release in a GPCR dependant manner [48]. *Ex vivo*, a reciprocal relationship has been described between LXA_4 and IL8 levels in the airways of patients with CF after antibiotic treatment [49].

LXA_4 is biosynthesised from Arachidonic Acid by the trans-cellular sequential action of Lipoxygenase enzymes [4]. LXA_4 concentration in bronchoalveolar lavage from the lungs of patients with CF has been variously reported as either significantly suppressed, or not significantly different from disease controls [47, 50]. The ability of platelets from patients with CF to contribute to LXA_4 generation was reported to be significantly reduced in a manner that was replicated in vitro by CFTR inhibition [51].

LXA_4 interacts with a G-Protein Coupled Receptor (GPCR) the Formyl Peptide Receptor 2 (FPR2) [52] located at the apex of bronchial epithelial cells [48].

LXA_4 has also been reported to be an intracellular ligand for the Aryl hydrocarbon receptor (AhR), a transcription factor which heterodimerizes with the AhR nuclear transporter protein upon ligand binding and interacts with dioxin response elements to regulate genes involved in xenobiotic metabolism [53, 54].

In airway epithelia LXA_4 has been described to stimulate a rise in intracellular calcium via adenylate cyclase activity and PKA activation [55]. Although further cell signalling pathways mediating the full range of LXA_4 effector functions have not yet been fully elucidated in airway epithelia, inferences can be made from studies conducted in other cell types, illustrated in Figure 2B. A survey of global epithelial gene expression in human enterocytes uncovered a host of genes under NFκB control whose transcription was reduced by LXA_4 [56]. In neutrophils, LXA_4 stimulation lead to an intracellular accumulation of Presqualene diphosphate (PSDP), which served to inhibit phospholipase D activation and potently inhibit superoxide anion generation [57]. In dendritic cells, Socs2 up regulation (a member of the suppressors of cytokine signalling family of proteins) played a central role in mediating the anti-inflammatory effects of LXA_4, in a manner that was dependent upon both AhR and FPR2 expression [54].

Role of Calcium in Regulating Airway Epithelial Ion Transport and Airway Surface Liquid Dynamics

Among many cellular processes, the intracellular calcium concentration modulates airway epithelial ion transport [1, 9], illustrated in Figure 1A. In particular, calcium activated chloride conductance (CaCC) has been widely reported in airway epithelia [12, 58-60] and plays a role in electrolyte secretion and airway surface hydration [61].

Two chloride channels have been reported to contribute to Calcium activated Chloride Conductance (CaCC) in the airways [62]. Bestrophin 1 (Best-1) shows ER localization [63], facilitates receptor-mediated Ca^{2+} signalling and probably serves as a Cl^- counter ion channel in the ER [64]. TMEM16A (also known as anoctamin 1) has been identified as the predominant component of CaCCs in the luminal airway epithelial membrane [64, 65].

The simplified schema describing two distinct pathways of Cl^- secretion; cAMP- activated pathway via CFTR; and, Ca^{2+}-activated chloride secretion is controversial. Extensive crosstalk between both chloride secretion pathways has now been revealed. The intracellular Cl^- concentration can directly interact with Ca^{2+} transport influx in ER via interaction between Best 1 (the ER Cl^- channel) and Orai1 and TRPC6 (the plasma membrane Ca^{2+} activated Cl^- channels) [64]. CFTR and TMEM16A may also interact directly or through scaffold proteins (see [66] for review). Other novel messengers, such as IP_3R binding protein (IRBIT) might link IP_3/Ca^{2+} signalling to CFTR activation although this has not yet been described in airway epithelial cells [67]. Purinergic receptors such as $P2Y_2$ and $P2Y_6$ couple the intracellular messengers, cAMP and Ca^{2+} [68, 69]. Likewise, calcium can affect the activity of enzymes that control the intracellular cAMP concentration such as adenylate cyclase and phophodiesterases [70]. Finally, cAMP influences proteins that control intracellular calcium concentration eg. SERCA and IP_3R [71].

Potassium channels participate in stabilization of the membrane potential necessary for Cl^- secretion and Na^+ absorption [9, 72-74]. Several K^+ channels have been identified and described in airway epithelial cells [75]. In particular, Ca^{2+}-activated K^+ channels have been described in airways epithelia including SK4, $K_{Ca}3.1$ and BK_{Ca} channels. Basolateral SK4 channels have been shown to play a role in Ca^{2+} dependent Cl⁻ secretion in cultured human bronchial epithelial cells [76]. Apical BK_{Ca} channels have recently been identified as being important for regulation the Periciliary layer volume in human bronchial epithelium [77].

Pseudomonas aeruginosa, a pathogen found in more than 80% of adult patients with CF, can induce an intracellular Ca^{2+} increase in airway epithelial cells [26, 27, 78, 79]. In fact, several studies have reported that *P. aeruginosa* virulence factors regulate ion transport. Rhamnolipids inhibit epithelial Na^+ absorption [80]. Flagellin has been shown to inhibit Na^+ absorption in murine tracheal epithelium [81] and to stimulate Cl⁻ secretion in human airway Calu-3 epithelial cells [82]. Homoserine lactones, small signalling molecules secreted by *P. aeruginosa*, activate CFTR and Cl⁻ secretion via stimulation of the cAMP pathway [83]. LPS inhibits Na^+ absorption via ENaC in mammalian alveolar epithelial cells [84]. More recently, we have demonstrated that LPS stimulates chloride secretion via the CFTR in human bronchial epithelium via stimulation of Ca^{2+} entry and release. We identified a role played by the TRP channel upon the aforementioned effect of LPS on $[Ca]_i$ [26]. Normal mucus clearance from the airways is mediated by a two phase liquid system that interfaces with beating cilia; the mucus phase and the periciliary liquid layer.The hydration of this Airway Surface Liquid (ASL) reflects the balance between sodium absorption, mediated principally by the Epithelial Sodium Channel (ENaC), and chloride secretion, mediated by the CFTR and the CaCC [5], illustrated in Figure 3A.

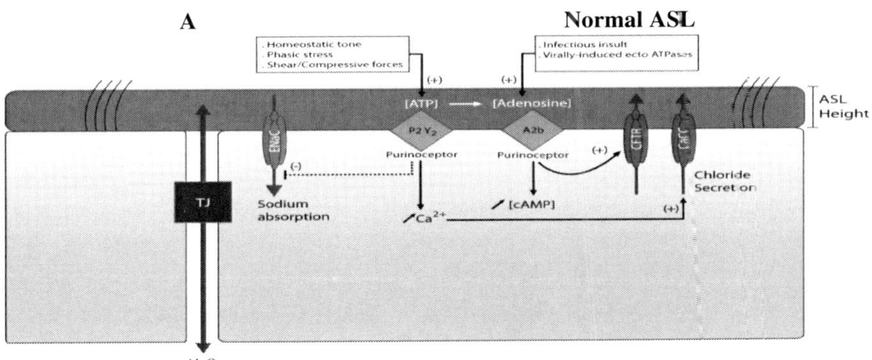

Figure 3. (Continued).

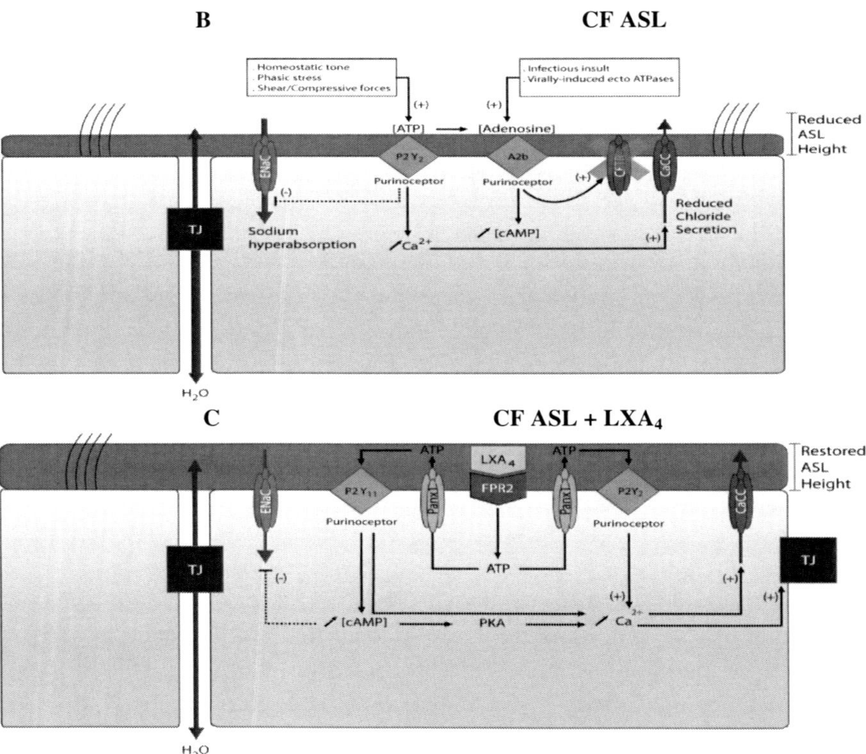

Figure 3. (A) Normal regulation of Airway Surface Liquid (ASL) height. (B) ASL height is reduced in CF. (C) Lipoxin A_4 restores ASL height on CF airway epithelial cells. (A) ASL volume is regulated by the surface liquid concentration of ATP and Adenosine. Under static conditions, airway epithelia release ATP onto the airway surface where it is converted to adenosine by cell surface exo-enzymes. Adenosine activates the A2b purino-receptor resulting in increased $[cAMP]_i$ and activation of CFTR regulation of ENaC and Chloride secretion. Phasic stress stimulates ATP release onto the airway surface where $[ATP]_{high}$ activates $P2Y_2$ purinoceptor resulting in ENaC inhibition and activation of chloride secretion via CFTR and CaCC. (B) In CF, loss of CFTR results in reduced ability to inhibit sodium absorption and secrete chloride. The ASL is more rapidly absorbed from airway surfaces and is unable to maintain a functional ASL height. The residual ASL height is relatively dependent on ATP signalling in CF, rendering the CF airway particularly vulnerable when viral infection results in the induction of ecto-ATP-ases. (C) LXA_4 stimulates a compensatory calcium-activated chloride secretory mechanism which overcomes the lack of CFTR mediated chloride transport and restores ASL height in vitro in CF epithelium. LXA_4 interacts with FPR2 to activate apical ATP release via Pannexin channels resulting in $P2Y_{11}$ purinoceptor stimulation. This generates increased $[Ca]_i$ and chloride secretion via CaCC. The result is restoration of the normal height and architecture of the ASL. Abbreviations: TJ, Tight Junction; ENaC, Epithelial Sodium channel; $P2Y_2$, P2Y Purinoceptor 2; A2b, Adenosine A2b receptor; CFTR, Cystic Fibrosis Transmembrane Conductance Regulator; CaCC, Calcium Activated Chloride channels; ASL, Airway Surface Liquid; $P2Y_{11}$, Purinergic Receptor P2.

As described earlier, cytosolic calcium concentration regulates the activity of the CaCC in the airway epithelial cell, and thus, contributes to the maintenance of the periciliary liquid height. The ciliary beat frequency (CBF), a key factor for the regulation of mucociliary transport, is also strongly regulated by Ca^{2+} [6-8].

ASL height is the integrated outcome of an intricately regulated biological system which is incompletely understood. The physiological regulation of ASL volume is signalled by the surface liquid concentration of ATP and Adenosine. Under static conditions, airway epithelia gradually release ATP onto the airway surface where it is converted to adenosine by cell surface exo-enzymes. This adenosine activates the A2b purino-receptor resulting in a rise in cAMP and activation of CFTR regulation of ENaC and Chloride secretion [85]. Phasic stresses stimulate relatively large increases in rates of ATP release onto the airway surface where a high concentration of ATP activates the $P2Y_2$ receptor and results in the inhibition of sodium absorption and activation of chloride secretion via CFTR and the CaCC [86].

In the normal airway both Adenosine and ATP signals are integrated and result in balanced Na^+ absorption and Cl^- secretion to maintain the dynamic volume of ASL on airway surfaces [5].

The electro-chemical driving force for the trans-epithelial movement of chloride is generated by the recycling of K^+ through potassium-selective ion channels in the basolateral membrane [87].

In CF, this system comes under pressure due to the loss of CFTR channel and regulatory activity and the consequent inability to both inhibit Na^+ absorption and initiate Cl^- secretion, illustrated in Figure 3B The ASL is more rapidly absorbed from airway surfaces and is unable to maintain a functional ASL height/volume under basal conditions.

This results in inefficient ciliary beating, mucus dehydration and adhesion of mucus plaques to airway surfaces. Evidence from animal experiments (including consideration of ciliary function and mucin hypersecretion) suggests that airway surface hydration is the most important variable in determining the efficiency of mucus clearance [5].

Airway cells can compensate for the absence of CFTR activity, especially during the motion phase, in part by increasing the release of ATP onto the airway surface, activating $P2Y_2$ receptors, inhibiting sodium absorption via ENaC and activating Chloride release via CaCC [86].

However, the CF airway is relatively dependent on ATP as its signalling mechanism rendering the CF airway particularly vulnerable when viral infection results in the induction of ecto-ATP-ases, upon which, the ASL volume can be seen to collapse [86].

One strategy to retard the progression of lung disease in CF is to pharmacologically restore the hydration of the ASL via stimulation of Calcium activated chloride conductance. Here we describe how endogenous mediators can impact upon ASL regulation.

Anti-Secretory Properties of Glucocorticoids

Steroid hormones including glucocorticoids have been reported to produce anti-secretory effects on various models of epithelial cells [88-90]. Several cellular mechanisms involved have been described and a central role has been attributed to the rapid non-genomic regulation of Potassium channel activity, intracellular pH and intracellular Calcium [89, 91-97]. More specifically glucocorticoids have been shown to regulate epithelial ion transport and the initial studies were performed on renal epithelial cells.

In airway epithelial cells, dexamethasone (even at low concentration) produces a concentration dependent, rapid decrease in intracellular calcium concentration which is sustained for at least 90 minutes [90, 98]. This suppressive effect on intracellular calcium signalling extends its reach to antagonise the intracellular calcium response produced by ATP upon activation of its purino-receptor in airway epithelial cells [98]. By suppressing intracellular calcium mobilisation, dexamethasone inhibits apical ATP induced chloride secretion in human bronchial epithelial cells. An inhibitory effect of fluticasone on cAMP-dependent chloride secretion has also been described [98].

The anti-secretory effect of glucocorticoids on airway epithelial cells and their antagonism of ATP driven, calcium activated chloride secretion, upon which CF airway epithelial cells are excessively reliant, is expected to have deleterious effects on ASL maintenance and regulation in CF. This negative effect on airway hydration may counterbalance the positive effect on nuclear regulation of inflammatory processes in CF and might be a contributory factor in the disappointing results obtained from trials of inhaled glucocorticoids as a therapeutic in CF.

Pro-Secretory Properties of Lipoxin A$_4$

LXA_4 produces distinct effects on intracellular Calcium concentration depending on the cell type. In airway epithelial cells LXA_4 stimulates an intracellular calcium increase [55]. Furthermore, LXA_4 stimulates a compensatory calcium-activated chloride secretory mechanism which overcomes the lack of CFTR mediated chloride transport and restores ASL height in vitro in CF epithelium [2]. LXA_4 interacts with its specific GPCR to generate a rapid release of calcium from its intracellular stores accompanied by extracellular calcium entry in CF cells.

This rise in intracellular calcium generates chloride secretion via the CaCC, which in turn restores the normal height and architecture of the ASL, replacing the dehydrated and disrupted appearance of the CF ASL prior to LXA_4 treatment [2], illustrated in Figure 3C. The increased intracellular calcium concentration generated by LXA_4 occurs via FPR2 stimulation, resulting in apical ATP release via Pannexin channels which results in $P2Y_{11}$ purino-receptor stimulation [99]. The unique ability of LXA_4 to stimulate increased intracellular calcium via stimulation of apical ATP secretion and purino-receptor activation yet activate a pro-resolution nuclear signalling pathway, and at the same time avoiding the classically detrimental effect of raised intracellular calcium upon pro-inflammatory signalling cascades, may be accounted for by one of two plausible explanations. Firstly, by employing the apical ATP release, purino-receptor stimulation circuit to achieve increased intracellular calcium concentration and thus Calcium activated chloride secretion, the FPR2 receptor coupling and signalling cascade may be spatially separated from the cellular micro-domain which experiences the peak of cytosolic calcium concentration increase. Secondly, FPR2 is an intriguing receptor which transduces very different intracellular responses in a ligand-specific fashion [100]. The ability to couple both the intracellular calcium increase, and independently, the pro-resolution signal transduction pathway may lie within the unique properties of the FPR2 receptor and its ability to couple diverse cytosolic binding partners. This question warrants further attention.

Pro-Resolution and Pro-Secretory: Therapeutic Perspective for Lipoxin A_4

LXA_4 augments airway epithelial innate defence by stimulating tight junction formation [101]. LXA_4 enhances Calcium activated Chloride secretion [55] and restores airway surface liquid height in CF airway epithelial cells [2]. LXA_4 has pro-resolution immune effector functions and in particular reduces pulmonary bacterial burden [47], limits neutrophil recruitment [102] and reduces IL8 secretion by airway epithelial cells [48]. There is an accumulating body of evidence suggesting that LXA_4 holds therapeutic promise for CF. One important hurdle in translating this potential had been the unstable nature of the lipid compound which is rapidly degraded by 15-PGDH [103]. To this end stable analogues have been synthesised [43, 104] and demonstrate comparable effects on ASL height in differentiated airway epithelial cell culture models (unpublished work).

Conclusion

While Glucocorticoids and Lipoxin A_4 both demonstrate anti-inflamatory properties in the airway, these two endogenous coumpounds differentially regulate ion transport and thus ASL hydration via opposite effects on intracellular calcium (Table 1).

Table 1. A summary of the divergent effects of Glucocorticoids and Lipoxin A4 upon airway hydration, intracellular calcium and inflammatory signalling

	Effect on Airway Hydration	Effect on Intracellular Calcium	Effect on Inflammatory Mediator Transcription
Glucocorticoids	ASL Dehydration	Decreased	Antagonises NFKB mediated pro-inflammatory mediator expression
Lipoxin A_4	ASL Hydration	Increased	Anti-inflammatory and Pro-resolution

References

[1]   Kunzelmann, K., Kubitz, R., Grolik, M., Warth, R., Greger, R. Small-conductance Cl- channels in HT29 cells: activation by Ca2+, hypotonic cell swelling and 8-Br-cGMP. *Pflugers Archiv: European journal of physiology.* 1992;421(2-3):238-46. Epub. 1992/06/01.

[2]   Verrière, V., Higgins, G., Al-Alawi, M., Costello, R. W., McNally, P., Chiron, R., et al. Lipoxin A$_4$ Stimulates Calcium-Activated Chloride Currents and Increases Airway Surface Liquid Height in Normal and Cystic Fibrosis Airway Epithelia. *PloS one.* 2012;7(5):e377 46.

[3]   Antigny, F., Norez, C., Becq, F., Vandebrouck, C. CFTR and Ca Signaling in Cystic Fibrosis. *Front Pharmacol.* 2011;2:67. Epub. 2011/11/03.

[4]   Serhan, C. N., Hamberg, M., Samuelsson, B. Lipoxins: novel series of biologically active compounds formed from arachidonic acid in human leukocytes. *Proceedings of the National Academy of Sciences of the US.* 1984;81(17):5335-9. Epub. 1984/09/01.

[5]   Boucher, R. C. Airway surface dehydration in cystic fibrosis: pathogenesis and therapy. *Annu. Rev. Med.* 2007;58:157-70. Epub. 2007/01/16.

[6]   Lansley, A. B., Sanderson, M. J., Dirksen, E. R. Control of the beat cycle of respiratory tract cilia by Ca2+ and cAMP. *The American journal of physiology.* 1992;263(2 Pt 1):L232-42. Epub. 1992/08/01.

[7]   Braiman, A., Zagoory, O., Priel, Z. PKA induces Ca2+ release and enhances ciliary beat frequency in a Ca2+-dependent and -independent manner. *The American journal of physiology.* 1998;275(3 Pt 1):C790-7. Epub. 1998/09/09.

[8]   Evans, J. H., Sanderson, M. J. Intracellular calcium oscillations regulate ciliary beat frequency of airway epithelial cells. *Cell calcium.* 1999; 26 (3-4):103-10. Epub. 1999/12/22.

[9]   Mall, M., Gonska, T., Thomas, J., Schreiber, R., Seydewitz, H. H., Kuehr, J., et al. Modulation of Ca2+-activated Cl- secretion by basolateral K+ channels in human normal and cystic fibrosis airway epithelia. *Pediatric research.* 2003;53(4):608-18. Epub. 2003/03/04.

[10]   Rebecchi, M. J., Pentyala, S. N. Structure, function, and control of phosphoinositide-specific phospholipase C. *Physiological reviews.* 2000; 80(4):1291-335. Epub. 2000/10/04.

[11] Petersen, C. C., Petersen, O. H., Berridge, M. J. The role of endoplasmic reticulum calcium pumps during cytosolic calcium spiking in pancreatic acinar cells. *The Journal of biological chemistry*. 1993;268(30):22262-4. Epub. 1993/10/25.

[12] Agnel, M., Vermat, T., Culouscou, J. M. Identification of three novel members of the calcium-dependent chloride channel (CaCC) family predominantly expressed in the digestive tract and trachea. *FEBS letters*. 1999;455(3):295-301. Epub. 1999/08/07.

[13] Liou, J., Kim, M. L., Heo, W. D., Jones, J. T., Myers, J. W., Ferrell, J. E., Jr., et al. STIM is a Ca2+ sensor essential for Ca2+-store-depletion-triggered Ca2+ influx. *Current biology: CB*. 2005;15(13):1235-41. Epub. 2005/07/12.

[14] Luik, R. M., Wu, M. M., Buchanan, J., Lewis, R. S. The elementary unit of store-operated Ca2+ entry: local activation of CRAC channels by STIM1 at ER-plasma membrane junctions. *The Journal of cell biology*. 2006;174(6):815-25. Epub. 2006/09/13.

[15] Roos, J., DiGregorio, P. J., Yeromin, A. V., Ohlsen, K., Lioudyno, M., Zhang, S., et al. STIM1, an essential and conserved component of store-operated Ca2+ channel function. *The Journal of cell biology*. 2005; 169 (3):435-45. Epub. 2005/05/04.

[16] Feske, S., Gwack, Y., Prakriya, M., Srikanth, S., Puppel, S. H., Tanasa, B., et al. A mutation in Orai1 causes immune deficiency by abrogating CRAC channel function. *Nature*. 2006;441(7090):179-85. Epub. 2006/04/04.

[17] Yeromin, A. V., Zhang, S. L., Jiang, W., Yu, Y., Safrina, O., Cahalan, M. D. Molecular identification of the CRAC channel by altered ion selectivity in a mutant of Orai. *Nature*. 2006;443(7108):226-9. Epub. 2006/08/22.

[18] Antigny, F., Norez, C., Dannhoffer, L., Bertrand, J., Raveau, D., Corbi, P., et al. Transient receptor potential canonical channel 6 links Ca2+ mishandling to cystic fibrosis transmembrane conductance regulator channel dysfunction in cystic fibrosis. *American journal of respiratory cell and molecular biology*. 2011;44(1):83-90. Epub. 2010/03/06.

[19] Huang, G. N., Zeng, W., Kim, J. Y., Yuan, J. P., Han, L., Muallem, S., et al. STIM1 carboxyl-terminus activates native SOC, I(crac) and TRPC1 channels. *Nature cell biology*. 2006;8(9):1003-10. Epub. 2006/08/15.

[20] Nilius, B., Prenen, J., Vennekens, R., Hoenderop, J. G., Bindels, R. J., Droogmans, G. Modulation of the epithelial calcium channel, ECaC, by intracellular Ca2+. *Cell calcium*. 2001;29(6):417-28. Epub. 2001/05/16.

[21] Ahmad, S., Ahmad, A., Dremina, E. S., Sharov, V. S., Guo, X., Jones, T. N., et al. Bcl-2 suppresses sarcoplasmic/endoplasmic reticulum Ca2+-ATPase expression in cystic fibrosis airways: role in oxidant-mediated cell death. *American journal of respiratory and critical care medicine.* 2009;179(9):816-26. Epub. 2009/02/10.

[22] Ji, Y., Lalli, M. J., Babu, G. J., Xu, Y., Kirkpatrick, D. L., Liu, L. H., et al. Disruption of a single copy of the SERCA2 gene results in altered Ca2+ homeostasis and cardiomyocyte function. *The Journal of biological chemistry.* 2000;275(48):38073-80. Epub. 2000/09/06.

[23] Ribeiro, C. M., Paradiso, A. M., Carew, M. A., Shears, S. B., Boucher, R. C. Cystic fibrosis airway epithelial Ca2+ i signaling: the mechanism for the larger agonist-mediated Ca2+ i signals in human cystic fibrosis airway epithelia. *The Journal of biological chemistry.* 2005;280 (11): 10202-9. Epub. 2005/01/14.

[24] Antigny, F., Norez, C., Cantereau, A., Becq, F., Vandebrouck, C. Abnormal spatial diffusion of Ca2+ in F508del-CFTR airway epithelial cells. *Respiratory research.* 2008;9:70. Epub. 2008/11/01.

[25] Norez, C., Antigny, F., Noel, S., Vandebrouck, C., Becq, F. A cystic fibrosis respiratory epithelial cell chronically treated by miglustat acquires a non-cystic fibrosis-like phenotype. *American journal of respiratory cell and molecular biology.* 2009;41(2):217-25. Epub. 2009/01/10.

[26] Buyck, J. M., Verriere, V., Benmahdi, R., Higgins, G., Guery, B., Matran, R., et al. P. aeruginosa LPS stimulates calcium signaling and chloride secretion via CFTR in human bronchial epithelial cells. *Journal of cystic fibrosis: official journal of the European Cystic Fibrosis Society.* 2012. Epub. 2012/07/20.

[27] Okuda, J., Hayashi, N., Arakawa, M., Minagawa, S., Gotoh, N. Type IV pilus protein PilA of Pseudomonas aeruginosa modulates calcium signaling through binding the calcium-modulating cyclophilin ligand. *Journal of infection and chemotherapy: official journal of the Japan Society of Chemotherapy.* 2012. Epub. 2012/12/26.

[28] Sel, S., Rost, B. R., Yildirim, A. O., Sel, B., Kalwa, H., Fehrenbach, H., et al. Loss of classical transient receptor potential 6 channel reduces allergic airway response. *Clinical and experimental allergy: journal of the British Society for Allergy and Clinical Immunology.* 2008; 38 (9): 1548-58. Epub. 2008/07/18.

[29] Balghi, H., Robert, R., Rappaz, B., Zhang, X., Wohlhuter-Haddad, A., Evagelidis, A., et al. Enhanced Ca2+ entry due to Orai1 plasma

membrane insertion increases IL-8 secretion by cystic fibrosis airways. *FASEB journal: official publication of the Federation of American Societies for Experimental Biology.* 2011;25(12):4274-91. Epub. 2011/08/30.

[30] Ciavardelli, D., D'Orazio, M., Pieroni, L., Consalvo, A., Rossi, C., Sacchetta, P., et al. Proteomic and ionomic profiling reveals significant alterations of protein expression and calcium homeostasis in cystic fibrosis cells. *Molecular bioSystems.* 2013;9(6):1117-26. Epub. 2013/04/24.

[31] Chan, H. C., Wu, W. L., So, S. C., Chung, Y. W., Tsang, L. L., Wang, X. F., et al. Modulation of the Ca(2+)-activated Cl(-) channel by 14-3-3epsilon. *Biochemical and biophysical research communications.* 2000; 270(2):581-7. Epub. 2001/02/07.

[32] Medzhitov, R., Preston-Hurlburt, P., Janeway, C. A., Jr. A human homologue of the Drosophila Toll protein signals activation of adaptive immunity. *Nature.* 1997;388(6640):394-7. Epub. 1997/07/24.

[33] Bezzerri, V., Borgatti, M., Finotti, A., Tamanini, A., Gambari, R., Cabrini, G. Mapping the transcriptional machinery of the IL-8 gene in human bronchial epithelial cells. *Journal of immunology* (Baltimore, Md: 1950). 2011;187(11):6069-81. Epub. 2011/10/28.

[34] Mukaida, N., Okamoto, S., Ishikawa, Y., Matsushima, K. Molecular mechanism of interleukin-8 gene expression. *J. Leukoc. Biol.* 1994;56 (5):554-8. Epub. 1994/11/01.

[35] Courtney, J. M., Ennis, M., Elborn, J. S. Cytokines and inflammatory mediators in cystic fibrosis. *Journal of cystic fibrosis : official journal of the European Cystic Fibrosis Society.* 2004;3(4):223-31. Epub. 2005/02/09.

[36] Kim, J. S., Okamoto, K., Rubin, B. K. Pulmonary function is negatively correlated with sputum inflammatory markers and cough clearability in subjects with cystic fibrosis but not those with chronic bronchitis. *Chest.* 2006;129(5):1148-54. Epub. 2006/05/11.

[37] Rottner, M., Freyssinet, J. M., Martinez, M. C. Mechanisms of the noxious inflammatory cycle in cystic fibrosis. *Respiratory research.* 2009;10:23. Epub. 2009/03/17.

[38] Rhen, T., Cidlowski, J. A. Antiinflammatory action of glucocorticoids--new mechanisms for old drugs. *N. Engl. J. Med.* 2005;353(16):1711-23. Epub. 2005/10/21.

[39] Nissen, R. M., Yamamoto, K. R. The glucocorticoid receptor inhibits NFkappaB by interfering with serine-2 phosphorylation of the RNA

polymerase II carboxy-terminal domain. *Genes Dev.* 2000;14(18):2314-29. Epub. 2000/09/20.

[40] Katharine, C. Oral steroids for long-term use in cystic fibrosis. *Cochrane Database of Systematic Reviews.* 2011;10.

[41] Balfour-Lynn Ian, M. W. K. Inhaled corticosteroids for cystic fibrosis. *Cochrane Database of Systematic Reviews.* 2009;1.

[42] Levy, B. D., Clish, C. B., Schmidt, B., Gronert, K., Serhan, C. N. Lipid mediator class switching during acute inflammation: signals in resolution. *Nat. Immunol.* 2001;2(7):612-9. Epub. 2001/06/29.

[43] Takano, T., Clish, C. B., Gronert, K., Petasis, N., Serhan, C. N. Neutrophil-mediated changes in vascular permeability are inhibited by topical application of aspirin-triggered 15-epi-lipoxin A4 and novel lipoxin B4 stable analogues. *The Journal of clinical investigation.* 1998; 101(4):819-26. Epub. 1998/03/21.

[44] Maddox, J. F., Serhan, C. N. Lipoxin A4 and B4 are potent stimuli for human monocyte migration and adhesion: selective inactivation by dehydrogenation and reduction. *The Journal of experimental medicine.* 1996;183(1):137-46. Epub. 1996/01/01.

[45] Godson, C., Mitchell, S., Harvey, K., Petasis, N. A., Hogg, N., Brady, H. R. Cutting edge: lipoxins rapidly stimulate nonphlogistic phagocytosis of apoptotic neutrophils by monocyte-derived macrophages. *Journal of immunology* (Baltimore, Md: 1950). 2000;164(4):1663-7. Epub. 2000/ 02/05.

[46] Special Issue: The lipoxins and the aspirin-triggered lipoxins. *Prostaglandins Leukot. Essent. Fatty Acids.* 2005;73(3-4):139-321. Epub. 2005/08/30.

[47] Karp, C. L., Flick, L. M., Park, K. W., Softic, S., Greer, T. M., Keledjian, R., et al. Defective lipoxin-mediated anti-inflammatory activity in the cystic fibrosis airway. *Nat. Immunol.* 2004;5(4):388-92. Epub. 2004/03/23.

[48] Bonnans, C., Gras, D., Chavis, C., Mainprice, B., Vachier, I., Godard, P., et al. Synthesis and anti-inflammatory effect of lipoxins in human airway epithelial cells. *Biomedicine and pharmacotherapy=Biomedecine and pharmacotherapie.* 2007;61(5):261-7. Epub. 2007/04/10.

[49] Chiron, R., Grumbach, Y. Y., Quynh, N. V., Verriere, V., Urbach, V. Lipoxin A(4) and interleukin-8 levels in cystic fibrosis sputum after antibiotherapy. *Journal of cystic fibrosis: official journal of the European Cystic Fibrosis Society.* 2008;7(6):463-8. Epub. 2008/06/11.

[50] Starosta, V., Ratjen, F., Rietschel, E., Paul, K., Griese, M. Anti-inflammatory cytokines in cystic fibrosis lung disease. *Eur. Respir. J.* 2006;28(3):581-7. Epub. 2006/06/30.

[51] Mattoscio, D., Evangelista, V., De Cristofaro, R., Recchiuti, A., Pandolfi, A., Di Silvestre, S., et al. Cystic fibrosis transmembrane conductance regulator (CFTR) expression in human platelets: impact on mediators and mechanisms of the inflammatory response. *FASEB journal: official publication of the Federation of American Societies for Experimental Biology.* 2010;24(10):3970-80. Epub. 2010/06/10.

[52] Gronert, K., Gewirtz, A., Madara, J. L., Serhan, C. N. Identification of a human enterocyte lipoxin A4 receptor that is regulated by interleukin (IL)-13 and interferon gamma and inhibits tumor necrosis factor alpha-induced IL-8 release. *The Journal of experimental medicine.* 1998;187 (8):1285-94. Epub. 1998/05/23.

[53] Schaldach, C. M., Riby, J., Bjeldanes, L. F. Lipoxin A4: a new class of ligand for the Ah receptor. *Biochemistry.* 1999;38(23):7594-600. Epub. 1999/06/09.

[54] Machado, F. S., Johndrow, J. E., Esper, L., Dias, A., Bafica, A., Serhan, C. N., et al. Anti-inflammatory actions of lipoxin A4 and aspirin-triggered lipoxin are SOCS-2 dependent. *Nat. Med.* 2006;12(3):330-4. Epub. 2006/01/18.

[55] Bonnans, C., Mainprice, B., Chanez, P., Bousquet, J., Urbach, V. Lipoxin A4 stimulates a cytosolic Ca2+ increase in human bronchial epithelium. *The Journal of biological chemistry.* 2003;278(13):10879-84. Epub. 2002/12/26.

[56] Gewirtz, A. T., Collier-Hyams, L. S., Young, A. N., Kucharzik, T., Guilford, W. J., Parkinson, J. F., et al. Lipoxin a4 analogs attenuate induction of intestinal epithelial proinflammatory gene expression and reduce the severity of dextran sodium sulfate-induced colitis. *Journal of immunology* (Baltimore, Md: 1950). 2002;168(10):5260-7. Epub. 2002/05/08.

[57] Levy, B. D., Fokin, V. V., Clark, J. M., Wakelam, M. J., Petasis, N. A., Serhan, C. N. Polyisoprenyl phosphate (PIPP) signaling regulates phospholipase D activity: a 'stop' signaling switch for aspirin-triggered lipoxin A4. *FASEB journal: official publication of the Federation of American Societies for Experimental Biology.* 1999;13(8):903-11. Epub. 1999/05/04.

[58] Jeulin, C., Guadagnini, R., Marano, F. Oxidant stress stimulates Ca2+-activated chloride channels in the apical activated membrane of cultured

nonciliated human nasal epithelial cells. *American journal of physiology Lung cellular and molecular physiology.* 2005;289(4):L636-46. Epub. 2005/09/09.

[59] Paradiso, A. M., Cheng, E. H., Boucher, R. C. Effects of bradykinin on intracellular calcium regulation in human ciliated airway epithelium. *The American journal of physiology.* 1991;261(2 Pt 1):L63-9. Epub. 1991/08/01.

[60] Grubb, B. R., Vick, R. N., Boucher, R. C. Hyperabsorption of Na+ and raised Ca(2+)-mediated Cl- secretion in nasal epithelia of CF mice. *The American journal of physiology.* 1994;266(5 Pt 1):C1478-83. Epub. 1994/05/01.

[61] Tarran, R. Regulation of airway surface liquid volume and mucus transport by active ion transport. *Proceedings of the American Thoracic Society.* 2004;1(1):42-6. Epub. 2005/08/23.

[62] Kunzelmann, K., Kongsuphol, P., Aldehni, F., Tian, Y., Ousingsawat, J., Warth, R., et al. Bestrophin and TMEM16-Ca(2+) activated Cl(-) channels with different functions. *Cell calcium.* 2009;46(4):233-41. Epub. 2009/09/29.

[63] Barro-Soria, R., Schreiber, R., Kunzelmann, K. Bestrophin 1 and 2 are components of the Ca(2+) activated Cl(-) conductance in mouse airways. *Biochimica et biophysica acta.* 2008;1783(10):1993-2000. Epub. 2008/07/26.

[64] Barro Soria, R., Spitzner, M., Schreiber, R., Kunzelmann, K. Bestrophin-1 enables Ca2+-activated Cl- conductance in epithelia. *The Journal of biological chemistry.* 2009;284(43):29405-12. Epub. 2006/09/28.

[65] Ousingsawat, J., Martins, J. R., Schreiber, R., Rock, J. R., Harfe, B. D., Kunzelmann, K. Loss of TMEM16A causes a defect in epithelial Ca2+-dependent chloride transport. *The Journal of biological chemistry.* 2009; 284(42):28698-703. Epub. 2009/08/15.

[66] Kunzelmann, K., Tian, Y., Martins, J. R., Faria, D., Kongsuphol, P., Ousingsawat, J., et al. Airway epithelial cells--functional links between CFTR and anoctamin dependent Cl- secretion. *The international journal of biochemistry and cell biology.* 2012;44(11):1897-900. Epub. 2012/06/20.

[67] Park, S., Shcheynikov, N., Hong, J. H., Zheng, C., Suh, S. H., Kawaai, K., et al. Irbit mediates synergy between ca(2+) and cAMP signaling pathways during epithelial transport in mice. *Gastroenterology.* 2013; 145(1):232-41. Epub. 2013/04/02.

[68] Faria, R. X., Reis, R. A., Casabulho, C. M., Alberto, A. V., de Farias, F. P., Henriques-Pons, A., et al. Pharmacological properties of a pore induced by raising intracellular Ca2+. *American journal of physiology Cell physiology.* 2009;297(1):C28-42. Epub. 2009/03/27.

[69] Schreiber, R., Kunzelmann, K. Purinergic P2Y6 receptors induce Ca2+ and CFTR dependent Cl- secretion in mouse trachea. *Cellular physiology and biochemistry: international journal of experimental cellular physiology, biochemistry, and pharmacology.* 2005;16(1-3):99-108. Epub. 2005/08/27.

[70] Namkung, W., Phuan, P. W., Verkman, A. S. TMEM16A inhibitors reveal TMEM16A as a minor component of calcium-activated chloride channel conductance in airway and intestinal epithelial cells. *The Journal of biological chemistry.* 2011;286(3):2365-74. Epub. 2010/ 11/ 19.

[71] Bruce, J. I., Straub, S. V., Yule, D. I. Crosstalk between cAMP and Ca2+ signaling in non-excitable cells. *Cell calcium.* 2003;34(6):431-44. Epub. 2003/10/24.

[72] Cowley, E. A., Linsdell, P. Characterization of basolateral K+ channels underlying anion secretion in the human airway cell line Calu-3. *The Journal of physiology.* 2002;538(Pt 3):747-57. Epub. 2002/02/05.

[73] Leroy, C., Dagenais, A., Berthiaume, Y., Brochiero, E. Molecular identity and function in transepithelial transport of K(ATP) channels in alveolar epithelial cells. *American journal of physiology Lung cellular and molecular physiology.* 2004;286(5):L1027-37. Epub. 2004/01/20.

[74] Leroy, C., Prive, A., Bourret, J. C., Berthiaume, Y., Ferraro, P., Brochiero, E. Regulation of ENaC and CFTR expression with K+ channel modulators and effect on fluid absorption across alveolar epithelial cells. *American journal of physiology Lung cellular and molecular physiology.* 2006;291(6):L1207-19. Epub. 2006/08/08.

[75] Bardou, O., Trinh, N. T., Brochiero, E. Molecular diversity and function of K+ channels in airway and alveolar epithelial cells. *American journal of physiology Lung cellular and molecular physiology.* 2009;296(2): L145-55. Epub. 2008/12/09.

[76] Bernard, K., Bogliolo, S., Soriani, O., Ehrenfeld, J. Modulation of calcium-dependent chloride secretion by basolateral SK4-like channels in a human bronchial cell line. *The Journal of membrane biology.* 2003; 196(1):15-31. Epub. 2004/01/16.

[77] Manzanares, D., Gonzalez, C., Ivonnet, P., Chen, R. S., Valencia-Gattas, M., Conner, G. E., et al. Functional apical large conductance, Ca2+-

activated, and voltage-dependent K+ channels are required for maintenance of airway surface liquid volume. *The Journal of biological chemistry.* 2011;286(22):19830-9. Epub. 2011/04/02.

[78] Ratner, A. J., Bryan, R., Weber, A., Nguyen, S., Barnes, D., Pitt, A., et al. Cystic fibrosis pathogens activate Ca2+-dependent mitogen-activated protein kinase signaling pathways in airway epithelial cells. *The Journal of biological chemistry.* 2001;276(22):19267-75. Epub. 2001/03/30.

[79] Jacob, T., Lee, R. J., Engel, J. N., Machen, T. E. Modulation of cytosolic Ca(2+) concentration in airway epithelial cells by Pseudomonas aeruginosa. *Infection and immunity.* 2002;70(11):6399-408. Epub. 2002/10/16.

[80] Graham, A., Steel, D. M., Wilson, R., Cole, P. J., Alton, E. W., Geddes, D. M. Effects of purified Pseudomonas rhamnolipids on bioelectric properties of sheep tracheal epithelium. *Experimental lung research.* 1993;19(1):77-89. Epub. 1993/01/01.

[81] Kunzelmann, K., Scheidt, K., Scharf, B., Ousingsawat, J., Schreiber, R., Wainwright, B., et al. Flagellin of Pseudomonas aeruginosa inhibits Na+ transport in airway epithelia. *FASEB journal: official publication of the Federation of American Societies for Experimental Biology.* 2006; 20 (3): 545-6. Epub. 2006/01/18.

[82] Illek, B., Fu, Z., Schwarzer, C., Banzon, T., Jalickee, S., Miller, S. S., et al. Flagellin-stimulated Cl- secretion and innate immune responses in airway epithelia: role for p. 38. *American journal of physiology Lung cellular and molecular physiology.* 2008;295(4):L531-42. Epub. 2008/07/29.

[83] Schwarzer, C., Wong, S., Shi, J., Matthes, E., Illek, B., Ianowski, J. P., et al. Pseudomonas aeruginosa Homoserine lactone activates store-operated cAMP and cystic fibrosis transmembrane regulator-dependent Cl- secretion by human airway epithelia. *The Journal of biological chemistry.* 2010;285(45):34850-63. Epub. 2010/08/27.

[84] Boncoeur, E., Tardif, V., Tessier, M. C., Morneau, F., Lavoie, J., Gendreau-Berthiaume, E., et al. Modulation of epithelial sodium channel activity by lipopolysaccharide in alveolar type II cells: involvement of purinergic signaling. *American journal of physiology Lung cellular and molecular physiology.* 2010;298(3):L417-26. Epub. 2009/12/17.

[85] Lazarowski, E. R., Tarran, R., Grubb, B. R., van Heusden, C. A., Okada, S., Boucher, R. C. Nucleotide release provides a mechanism for airway surface liquid homeostasis. *The Journal of biological chemistry.* 2004; 279 (35):36855-64. Epub. 2004/06/24.

[86]   Tarran, R., Button, B., Picher, M., Paradiso, A. M., Ribeiro, C. M., Lazarowski, E. R., et al. Normal and cystic fibrosis airway surface liquid homeostasis. The effects of phasic shear stress and viral infections. *The Journal of biological chemistry*. 2005;280(42):35751-9. Epub. 2005/08/10.

[87]   Saint-Criq, V., Rapetti-Mauss, R., Yusef, Y. R., Harvey, B. J. Estrogen regulation of epithelial ion transport: Implications in health and disease. *Steroids*. 2012;77(10):918-23. Epub. 2012/03/14.

[88]   Urbach, V., Harvey, B. J. Rapid and non-genomic reduction of intracellular [Ca(2+)] induced by aldosterone in human bronchial epithelium. *The Journal of physiology*. 2001;537(Pt 1):267-75. Epub. 2001/11/17.

[89]   Harvey, B. J., Doolan, C. M., Condliffe, S. B., Renard, C., Alzamora, R., Urbach, V. Non-genomic convergent and divergent signalling of rapid responses to aldosterone and estradiol in mammalian colon. *Steroids*. 2002;67(6):483-91. Epub. 2002/04/19.

[90]   Urbach, V., Walsh, D. E., Mainprice, B., Bousquet, J., Harvey, B. J. Rapid non-genomic inhibition of ATP-induced Cl- secretion by dexamethasone in human bronchial epithelium. *The Journal of physiology*. 2002;545(Pt 3):869-78. Epub. 2002/12/17.

[91]   Condliffe, S. B., Doolan, C. M., Harvey, B. J. 17beta-oestradiol acutely regulates Cl- secretion in rat distal colonic epithelium. *The Journal of physiology*. 2001;530(Pt 1):47-54. Epub. 2001/01/04.

[92]   Doolan, C. M., Condliffe, S. B., Harvey, B. J. Rapid non-genomic activation of cytosolic cyclic AMP-dependent protein kinase activity and [Ca(2+)](i) by 17beta-oestradiol in female rat distal colon. *British journal of pharmacology*. 2000;129(7):1375-86. Epub. 2000/04/01.

[93]   Maguire, D., MacNamara, B., Cuffe, J. E., Winter, D., Doolan, C. M., Urbach, V., et al. Rapid responses to aldosterone in human distal colon. *Steroids*. 1999;64(1-2):51-63. Epub. 1999/05/14.

[94]   Doolan, C. M., O'Sullivan, G. C., Harvey, B. J. Rapid effects of corticosteroids on cytosolic protein kinase C and intracellular calcium concentration in human distal colon. *Molecular and cellular endocrinology*. 1998;138(1-2):71-9. Epub. 1998/07/31.

[95]   Verriere, V. A., Hynes, D., Faherty, S., Devaney, J., Bousquet, J., Harvey, B. J., et al. Rapid effects of dexamethasone on intracellular pH and Na+/H+ exchanger activity in human bronchial epithelial cells. *The Journal of biological chemistry*. 2005;280(43):35807-14. Epub. 2005/07/26.

[96] O'Mahony, F., Alzamora, R., Chung, H. L., Thomas, W., Harvey, B. J. Genomic priming of the antisecretory response to estrogen in rat distal colon throughout the estrous cycle. *Molecular endocrinology* (Baltimore, Md). 2009;23(11):1885-99. Epub. 2009/10/23.

[97] Rapetti-Mauss, R., O'Mahony, F., Sepulveda, F. V., Urbach, V., Harvey, B. J. Oestrogen promotes KCNQ1 potassium channel endocytosis and postendocytic trafficking in colonic epithelium. *The Journal of physiology*. 2013;591(Pt 11):2813-31. Epub. 2013/03/27.

[98] Urbach, V., Verriere, V., Grumbach, Y., Bousquet, J., Harvey, B. J. Rapid anti-secretory effects of glucocorticoids in human airway epithelium. *Steroids*. 2006;71(4):323-8. Epub. 2005/11/22.

[99] Higgins, G. B. P., Perriere, M., Al-Alawi, M., Costello, R., Verriere, V., Chiron, R., McNally, P., Harvey, B. J., and Urbach, V. Activation of P2RY and ATP release by LXA$_4$ restores ASL and epithelial repair in cystic fibrosis. *American journal of respiratory cell and molecular biology*. 2013; Under Revision.

[100] Cattaneo, F., Parisi, M., Ammendola, R. Distinct signaling cascades elicited by different formyl Peptide receptor 2 (FPR2) agonists. *International journal of molecular sciences*. 2013;14(4):7193-230. Epub. 2013/04/04.

[101] Grumbach, Y., Quynh, N. V., Chiron, R., Urbach, V. LXA4 stimulates ZO-1 expression and transepithelial electrical resistance in human airway epithelial (16HBE14o-) cells. *American journal of physiology Lung cellular and molecular physiology*. 2009;296(1):L101-8. Epub. 2008/10/14.

[102] Colgan, S. P., Serhan, C. N., Parkos, C. A., Delp-Archer, C., Madara, J. L. Lipoxin A4 modulates transmigration of human neutrophils across intestinal epithelial monolayers. *The Journal of clinical investigation*. 1993;92(1):75-82. Epub. 1993/07/01.

[103] Clish, C. B., Levy, B. D., Chiang, N., Tai, H. H., Serhan, C. N. Oxidoreductases in lipoxin A4 metabolic inactivation: a novel role for 15-onoprostaglandin 13-reductase/leukotriene B4 12-hydroxydehydrogenase in inflammation. *The Journal of biological chemistry*. 2000;275 (33):25372-80. Epub. 2000/06/06.

[104] Borgeson, E., Docherty, N. G., Murphy, M., Rodgers, K., Ryan, A., O'Sullivan, T. P., et al. Lipoxin A(4) and benzo-lipoxin A(4) attenuate experimental renal fibrosis. *FASEB journal: official publication of the Federation of American Societies for Experimental Biology*. 2011;25(9): 2967-79. Epub. 2011/06/02.

In: New Developments in Calcium … ISBN: 978-1-62948-601-7
Editor: Masayoshi Yamaguchi © 2014 Nova Science Publishers, Inc.

Chapter III

Excitation–Transcription Coupling: A Quantitative Approach

Evgeny Kobrinsky[1] and Nikolai M. Soldatov[1,2]
[1]National Institute on Aging, National Institutesof Health,
Baltimore, MD, US
[2]Humgenex, Inc., Kensington, MD, US

Abstract

The function of neurons, cardiomyocytes and other excitable cells is affected by hormones and the changing patterns of the plasma membrane electrical activity. Voltage-gated $Ca_v1.2$ calcium channels and cAMP/G protein-coupled receptors are the two major pathways communicating external excitation to the cell nucleus to regulate transcription in the process named excitation–transcription coupling. The exact signaling events involved in excitation–transcription coupling are poorly understood. In this review we discuss the recent progress made in the research of transcription regulation by the $Ca_v1.2$ calcium channels. We will focus on the use of wavelet transform, the new powerful statistical approach to FRET image analysis, which revealed the spatio-temporal domain organization of the nuclear signaling associated with excitation–transcription coupling.

Introduction

Activity-dependent gene expression is critical for proper functioning of many cells, above all neuronal and muscle cells. The key process essential for activity-dependent gene expression is excitation-transcription (E-T) coupling (Dolmetsch et al., 2001, Bers, 2011).

Changes of transmembrane potential in neuronal and muscle cells affect their functional state, e.g., affect frequency of action potentials (AP). The change in dynamics of electrical activity results in changing the frequency of opening of the underlying voltage-gated calcium channels. The dynamics of opening of voltage gated $Ca_v1.2$ calcium channels, in turn, modulates the intracellular calcium signaling (Bers, 2008).

Transcriptional activity in neuronal and muscle cells is under control of a variety of signaling pathways. Some of the most important pathways, especially the cAMP-response-element-binding protein (CREB) dependent pathways, are regulated by the activity of second messengers - cAMP and calcium (Flavell, S. W., and Greenberg, M. E. (2008), Sands W. A., and Palmer T.M. (2008)). Coupling between the plasma membrane depolarization and gene expression is critical for the long-term changes of excitable cells, such as synaptic development and neuronal plasticity (Greer and Greenberg (2008)) in neurons or development of hypertrophy in cardiac muscle (Domínguez-Rodríguez et al., 2012).

Despite many unknowns, the recent progress in this field revealed a unique role of the L-type voltage-gated Ca^{2+} channel ($Ca_v1.2$) as the voltage sensor in the transmission of the Ca^{2+} signal to the nucleus in E-T coupling (Figure 1) Why $Ca_v1.2$ plays this important role and what are the major signaling characteristics of the $Ca_v1.2$-mediated E-T coupling are important questions to be asked. In recent years, a few quantitative experimental approaches were developed to address these questions.

Integrative Approach to the Study of Excitation-Transcription Coupling

In the experimental study of E-T coupling, the changes of transmembrane potential are typically compared with the changes of the integral response in phosphorylation of CREB in the nuclei as a measure of CREB-dependent transcriptional activation.

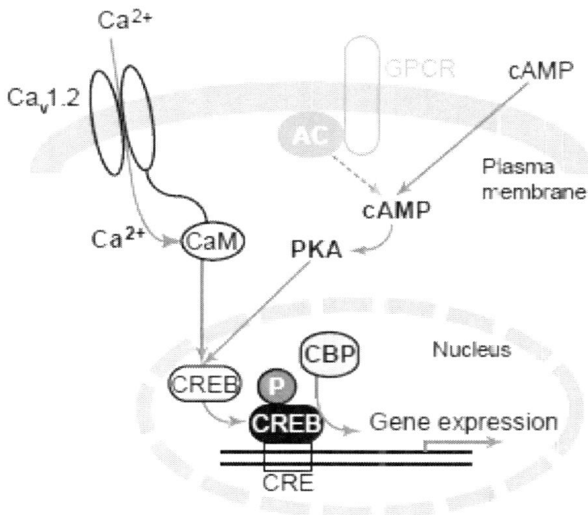

Reproduced with permission from *FASEB J.* (Kobrinsky et al., 2011).

Figure 1. Ca$_v$1.2- and cAMP/G protein-coupled cell signaling pathways mediating CREB-dependent transcriptional activation. Activation of Cav1.2 channels results in CaM and cAMP-dependent CREB phosphorylation of Ser-133. This results in CREB binding to CRE through the interaction with CBP, leading to activation of gene transcription. All these signaling events constitute E-T coupling.

The effectiveness of excitation–response coupling depends on the sensitivity of the signaling to the changes of transmembrane voltage. Thus, to measure the effectiveness of the coupling, one has to compare changes of transmembrane voltage with the strength of the induced signal.

This can be successfully achieved under patch clamp conditions in the live cell expressing probes for transcriptional activation. Phosphorylation of CREB inside the nuclei can be used an integrated parameter for measuring the strength of transcriptional signaling in CREB-dependent transcription activation. Because PKA-mediated phosphorylation of CREB on Ser133 is essential for activation of transcriptional response to second messenger cAMP (Sands and Palmer, 2008), in earlier experiments (Wheeler et al., 2008) a fluorescently labeled antibody to CREB Ser133 was used. In this experimental approach, cells were depolarized with different concentrations of K$^+$ and fixed in ice-cold 4% formaldehyde in phosphate buffer supplemented with 4% sucrose. The fixed cells were sequentially incubated with a primary anti-pCREB antibody and then with a secondary ALEXA-488 antibody.

The intensity of the fluorescent signal over the whole nuclear region indicated the level of CREB activation. The dependence of intensity of the fluorescent signal on changes of transmembrane potential was measured in these experiments, and the initial slope of the curve of the level of CREB phosphorylation over time of cell depolarization was estimated. This parameter (the slope) was named the "CREB signaling strength" (Wheeler et al., 2008).

Then the signaling strength was compared with the voltage-dependence of other cellular processes including, e.g., contraction in cardiac myocytes or secretion in neuroendocrine cells.

Because this is an integrated approach, it provides net characterization of transcription under assumption that the signaling process in the nuclei is homogeneous in space and time.

However, in reality a mammalian cell nucleus is structurally and functionally complex and contains morphologically distinct chromatin domains and numerous protein subcompartments. This non-uniform nuclear organization results in a very discrete spatial heterogeneity of the nuclear signaling. Indeed, it is known that chromosomes occupy discrete, nonrandom positions inside a nucleus (Parada et al., 2004). The position of chromosomes and special organization of actively transcribed genes convey the E-T coupling a natural heterogeneity in spatial structures of actively transcribed nuclear microdomains (Maharana et al., 2012).

Thus, gene expression is organized in spatially heterogeneous subnuclear domains (Bergman et al., 1995). It is generally accepted that the spatial organization of these nuclear compartments is inherently connected to their role in gene expression and cell regulation (Trinkle-Mulcahy and Lamond, 2008). The dynamic organization of these transcriptional compartments is coupled to functional nuclear spatial organization in a transcription-dependent manner. Because of this tremendous complexity, an integral approach can be misleading due to the transient and dynamic nature of transcriptional regulation.

Therefore, to investigate the activity of transcription and transcriptional regulation one should identify and monitor the spatio-temporal activity of transcriptional compartments organized in signaling microdomains (Kobrinsky E. et al., 2011).

Elucidation of CREB Signaling Microdomains by Wavelet Transform Analysis

In cell microscopy imaging of transcriptional activation, each pixel of the recorded nuclear image contains information regarding the activity of signaling events. To address the heterogeneity of signaling events, we have to identify signaling domains, i.e., pixels or groups of pixels representing specific activity. The appropriate statistical methods should be used in the identification of significant changes in signaling over background activity.

To reach this important goal, we proposed to apply wavelet transform to the analysis of E-T coupling in the live cell (Kobrinsky et al., 2005; Mager et al., 2007). Our experimental setting included FRET microscopy imaging of CREB-dependent transcriptional activation in live cells expressing recombinant (COS1 or HEK293) or endogenous (cardiac myocytes) $Ca_v1.2$ combined with patch clamp. This setting permitted effective control over activation of CREB-dependent transcription by stimulation of the inward Ca^{2+} current through $Ca_v1.2$ as well as by external application of cAMP in various combinations. To correlate signaling events inside the nucleus with the activity of $Ca_v1.2$ in the plasma membrane, we simultaneously recorded two parameters: excitability of the plasma membrane labeled with voltage sensing dye di-8 ANEPPS (Satin et al., 2008) and intranuclear CREB-dependent transcriptional signaling measured as FRET signal of endogenously expressed interacting domains of CREB and CREB binding protein (CBP) fused to EYFP and ECFP, respectively (Kobrinsky E., et al., 2003).

The wavelet transform is an extension of Fourier analysis but colocalizes in both the space and frequency domains to permit assessment of changes in spatial frequency content of obtained 2D images. Thus, this mathematical technique allows us to analyze a signal over several different frequencies across the entire signal (Kobrinsky et al., 2005, Mager et al., 2007). To do this, we compare the real signal to another signal called a *wavelet* (see Figure 2) across the entire length of the nucleus.

For each local region we get a measure of how well the wavelet correlates to the signal it is compared to, called a *scale coefficient*. We then shift the wavelet to the next region in the nucleus and measure another coefficient, and so on until we cover the whole signaling event with the chosen wavelet. Then, we change the frequency of the wavelet by stretching or contracting it, an operation known as *scaling* the wavelet.

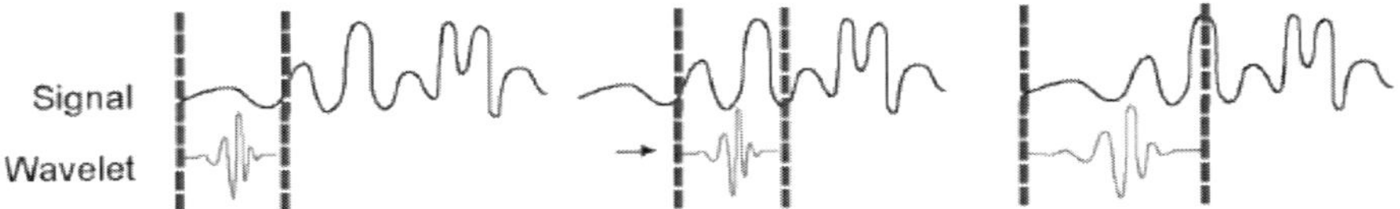

Figure 2. Comparison of the signal to the wavelet function. Shown are analyzing wavelet shifted over the analyzed signal (*center*) and scaling (stretching) wavelet (*right*) to a higher scale.

We can now repeat the comparison of the signal and the wavelet with the newly scaled wavelet, to see how the signal compares to the wavelet at a different frequency. By repeating this over a number of scales, we can build a graph called a *scalogram*. That shows how strongly a signal resembles the wavelet it is compared to within a range of frequencies. In other words, we can split the signal into component frequencies while retaining how the frequencies are localized within the signal. Using this mathematical technique permitting simultaneously global and local analysis of the signaling event, we can find out whether in a given time series signal X there are drastic changes in frequency and amplitude, and how the two time series are related in frequency and phase. To implement this powerful analytical method to the study of transcriptional activation, we developed continuous 2D wavelet transform as a deconvolution algorithm for FRET microscopy image analysis of CREB-dependent transcriptional activation (Mager et al., 2007). Naturally, this statistical approach is particularly useful for the identification of *heterogeneity* in a signal as it can easily find where the pattern (i.e., frequency) of a signal changes. We used this method to identify localized regions of increased FRET intensity of the CREB transcriptional probes (see below) within the nucleus, which we call microdomains (although in fact they are submicro-domains). The detailed mathematical apparatus and description of wavelet analysis of FRET microscopy images were described in a series of our papers (Kobrinsky et al., 2005, Mager et al., 2007, Kobrinsky et al., 2011).

The images are analyzed as coefficient matrices while their differences with controls are used to initially identify the potential microdomains that can be accepted or rejected based on the subsequent statistical comparison. Then one can compare spatio-temporal pattern of signaling activity in already identified signaling domains. Our findings supported the existence of a significant heterogeneity of CREB-dependent transcription in both native and recombinant cell systems, where wavelet transform analysis identified transient and stable discrete CREB-dependent transcriptional signaling domains.

To understand how synchronous and coherent are the events in the plasma membrane and the identified spatially different segregated nuclear microdomains, we applied time-series statistical analysis. Wavelet transform coherence is a well established technique for time series analysis (Grinsted et al., 2004). First, 1D continuous wavelet transform analysis (Kobrinsky, et al., 2005) was used to expand time series of plasma membrane signals and signals within an identified intranuclear spatial domain into a time and frequency space with Morlet transform. This wavelet function is a good choice for feature extraction purposes because it is reasonably localized in time and frequency (Grinsted et al., 2004). This approach provided an opportunity to detect areas in times series with significant transient responses. Then wavelet transform coherence (WTC) was used to analyze the local correlation between two simultaneously recorded processes: membrane excitability and nuclear transcriptional signaling. 1D continuous wavelet transform analysis was applied to the plasma membrane and nuclear signals. By using WTC we quantitatively measured the extent of the coherence between excitability of the plasma membrane and intranuclear transcriptional activation.

The two discussed quantitative approaches in studying E-T coupling highlight different aspects of transcriptional regulation. The first approach is based on the recording of integral fluorescent signal from the nuclei. This methodology was used to estimate the extent and strength of the coupling between the changes in transmembrane potential and CREB phopshorylation inside the nuclei. The second approach, developed in our laboratory, is based on collecting FRET reporter signals from nuclei and sophisticated analysis with the wavelet method aimed at detecting spatially segregated signaling microdomains. In addition, WTC analysis helps to estimate the level of synchronicity and coherence between the excitability of the plasma membrane and the induced CREB-dependent transcriptional signaling events. The progress in research of E-T coupling by the wavelet transform quantitative approach is shown in the next chapter.

Excitation-Transcription Coupling Signaling

During action potentials, the L-type $Ca_v1.2$ calcium channels couple membrane depolarization to transient increase in intracellular free calcium concentration, which in part is involved in the regulation of CREB-dependent transcriptional activity. However, other types of voltage-gated calcium channels (e.g., the N and P/Q type Ca_v2 channels) are much less effective in E-T coupling (Wheeler et al., 2012). The authors suggested three major reasons for the superiority of the $Ca_v1.2$ channel in the E-T coupling. First, $Ca_v1.2$ has gating advantage at low voltages. Indeed, the probability of opening for $Ca_v1.2$ is higher at a relatively moderate depolarization (> -60 mV), thus providing lower threshold for maximal signal strength response. Second, Ca^{2+} transients resulting from the activation of $Ca_v1.2$ appear to signal to CREB ~10 times more efficiently than Ca^{2+} from Ca_v2 channels. To elicit a CREB-dependent transcriptional response, Ca^{2+} entering through Ca_v2 channels must act at a considerably greater distance from the site of Ca^{2+} entry (~1 μm), while $Ca_v1.2$ channels signal in the vicinity of the pore by activating CaMKII. Finally, it is critical for effective E-T coupling that the specific supramolecular organization of $Ca_v1.2$ channels in signaling complexes with CaMKII and other downstream signaling pathway players, which explains effective local Ca^{2+} signaling in the vicinity of $Ca_v1.2$.

The structure-functional features of $Ca_v1.2$ that provide for this distinct behavior were found in our earlier studies of voltage-gated mobility of the $Ca_v1.2$ pore-forming α_{1C} subunit C-terminal tail (Kobrinsky et al., 2003). This 663-amino acid peptide is located in the cytoplasm and contains a single calmodulin (CaM) molecule shared between two adjacent binding sites that correspond to its apo– and Ca^{2+}-bound states (for review, see Halling et al. (2005)). FRET experiments under patch clamp conditions in the live COS1 cell expressing fully functional endogenous $Ca_v1.2$ containing ECFP/EYFP-labeled α_{1C} N- and C-termini showed large voltage-gated rearrangements of this C-terminal tail. Inhibition of the mobility of the C-tail by its anchoring in the membrane did not decrease the calcium current – in fact, it was much larger due to inhibition of the channel spontaneous inactivation. However, the anchoring of the α_{1C} subunit C-terminal tail completely inhibited activation of CREB-dependent transcription in spite of the natural proximity of CaMKII to the conducting $Ca_v1.2$ pore. Release of the tail fully restored the voltage-gated mobility and transcriptional activation (Kobrinsky et al., 2003). Thus, neither the large inward Ca^{2+} current nor the massive intracellular Ca^{2+} release lead to CREB-dependent transcription activation, which is entirely due to the voltage-gated shuttling of CaM as Ca^{2+} carrier between the pore and the target by the mobile α_{1C} subunit C-tail, which underlies activation and inactivation of $Ca_v1.2$ (Soldatov, 2003, Soldatov, 2012). Data obtained elsewhere (Bers, 2011, Wheeler et al., 2008, Wheeler et al., 2012) showed that calmodulin and CaMKII are the most important co-players in E-T coupling. CaMKII acting near the $Ca_V1.2$ channel couples, decodes and amplifies Ca^{2+} signal to CREB-dependent transcriptional (CDT) activation by reacting to the frequency of the $Ca_V1.2$ openings rather than to the net Ca^{2+} flux.

However, CaMKII involvement in CREB-dependent transcriptional activation is not obsolete. Our study based on wavelet analysis showed (Kobrinsky et al., 2011) that a significant fraction of nuclear microdomains of CREB-dependent transcriptional activity in COS1 cells expressing recombinant $Ca_V1.2$ calcium channels as well as in native cardiac myocytes were not sensitive to CaMKII inhibitors. We observed four major types of intranuclear microdomains involved in CDT signaling underlying its architecture. The majority of signaling domains in a COS1 cell model responded to both cAMP external application and $Ca_V1.2$ channel stimulation, showing sustained (65%) and transient (15%) activity. These domains constitute changes in discrete FRET signals that essentially reflect the average nuclear FRET changes, measured by the conventional integral approach. However, approximately 20% of stable CDT microdomains showed a differential response to either cAMP or to the $Ca_V1.2$ current stimulation. The CaMKII inhibitor KN93 completely abolished cAMP-induced CDT signaling.

We found that cardiac E-T coupling signaling is also organized in stable and transient nuclear microdomains activated by calcium channel current, cAMP, or both of these stimuli. Our data showed that in neonatal rat cardiac myocytes, the CREB-dependent E-T coupling is partially executed through CaMKII-independent pathways that have not been observed in neurons. About 1/3 of basal CREB/CBP interactions in heart cells depend on $Ca_V1.2$ channel activity but are not affected by the inhibition of CaMKII and therefore do not depend on CaMKII phosphorylation. The similarity of the CDT spatiotemporal organization in cardiac myocytes and recombinant COS1 cells suggests that characteristic architecture of CDT signaling is inherited in a wide range of different cells. Moreover, it can be rendered even in cells naturally deprived of $Ca_V1.2$, such as COS1 cells.

In cardiac myocytes we also detected microdomains showing periodic CDT activity. Periodic oscillations in gene expression were observed in other types of cells (Klevecz and Li, 2007). Pacing-induced reduction in the CREB level was also observed in heart cells (Patberg and Rosen, 2003, Rosen and Cohen, 2006). Periodic heart beats are known to result from the periodic Ca^{2+} current activation of $Ca_V1.2$ channels in cardiac cells. The same periodic Ca^{2+} influx provided the basis for frequency modulation of transcriptional signaling (Özgen and Rosen, 2009). Frequency-dependent CDT signaling that we observed in spontaneously contracting rat neonatal cardiomyocytes matched contractions and were sensitive to DHP calcium channel blocker.

Comparison of contraction activity and average FRET intensity of a single CDT transient microdomain in time suggest that the contraction reflecting the influx through $Ca_V1.2$ and the transcription activity in these cells are linked on both small and large time scales.

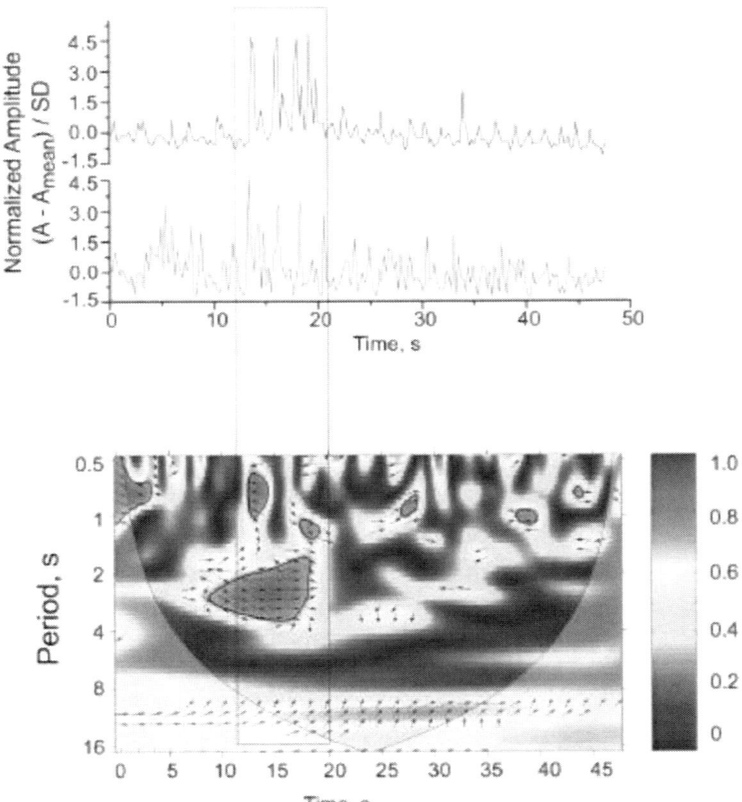

Reproduced with permission from *FASEB J.* (Kobrinsky et al., 2011).

Figure 3. Coherence time-series analysis of spontaneous contractility and CDT signaling activity in neonatal rat cardiac myocytes. *A)* Simultaneous recording of contraction and FRET between the KID and KIX probes within an identified transient nuclear CDT microdomain (~0.3 μm^2 in area). B) Wavelet coherence between spontaneous contractile activity and FRET signal. Regions outlined in black represent statistically significant areas based on Monte Carlo analysis (*P*<0.05). Arrows represent phase difference between contractile and transcriptional signaling data. Arrows pointing to the right are in phase, arrows pointing to the left are in antiphase.

Indeed, WTC confirmed that contraction and transcription activity exhibit a statistically significant phase-locked relationship (Figure 3).

We also combined confocal microscopy with 3D wavelet and cross-correlation analyses for visualization of CDT microdomains. The software has been written in collaboration with the YAWTB team to compute the 3D continuous wavelet transform to elucidate the 3D architecture of CDT in nuclei (Figure 4). In this study we performed a 3D FRET recording by z-scanning (24 z slices) before and during cAMP application. Each z slice of FRET image was subject to 2D-CWT analysis, and then a 3D reconstruction of nuclear CDT signaling microdomains was performed with C-imaging software. This 3D reconstruction revealed a spatial intranuclear organization of cAMP-dependent microdomains of CREB and CBP interaction. Some of these domains are on the periphery of the nucleus, some are localized closer to the center, but confirming specificity of our CWT analysis, no CDT signaling was observed in the nucleolar region.

CREB and CBP interactions in transient domains may play the most important part in the pacing (frequency)-dependent regulation of transcriptional activity in the heart (Özgen and Rosen, 2009). Transient responses are known to be associated with the stimulation of CREB activity by phosphorylation of serine-133 as well as by the activity of serine/threonine protein phosphatase (Koch et al., 2003) with additional contribution of phosphorylation of serine-142 and 143 of CREB (Kornhauser et al., 2002). Phosphatase activity as well as phosphorylation may provide a negative feedback in Ca^{2+}-dependent regulation of CREB activity and CREB/CBP interaction. It is clear that dynamics of phosphorylation/dephosphorylation of CREB regulatory sites is crucial for the appearance of transient microdomains of CREB and CBP interaction underlying E-T coupling signaling in heart cells. The wavelet transform approach to E-T coupling may facilitate understanding of these fine mechanisms.

Conclusion

CREB-dependent signaling is the major pathway in E-T coupling. Recent progress in E-T coupling investigation has been achieved by using quantitative methods. Voltage-gated $Ca_V1.2$ channels have a privileged role in the initiation of E-T coupling process. A proper conformation of C-terminal tail of the Ca$_V$1.2 α_1 subunit, CaM, CaMKII and local calcium signaling are important factors in transmitting activation signals to the nuclei.

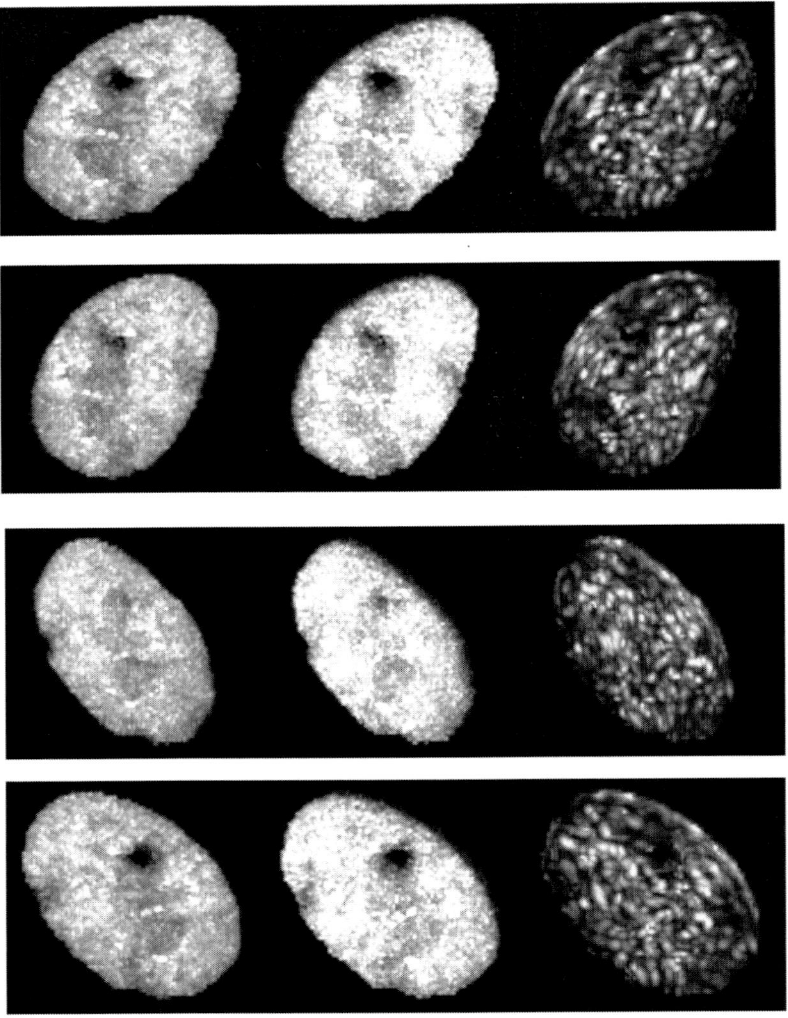

Figure 4. 3D reconstruction of cAMP activated microdomains of CREB and CBP interaction in the nucleus of a representative COS1 cells, viewed from different angles. *z*-Scanning was performed on Leica TCS SP confocal microscope. *Left:* FRET signal before application of cAMP. *Middle*: FRET signal after 10 min of cAMP application (0.25 mM) *Right*: 3D reconstruction of FRET signal calculated as the difference between control and cAMP-induced FRET. Microdomains of CREB and CBP interaction are clearly seen in the nuclei, except nucleoli region.

Intranuclear CREB-dependent transcriptional signaling is organized in spatially segregated microdomains. The majority of these domains in the nuclei of heart cells are stable and respond to both cAMP and $Ca_V1.2$ stimulation. The transient domains exhibit periodic behavior and may serve the frequency modulation of CREB-dependent transcriptional signaling in the heart, providing the basis of frequency-dependent regulation of transcriptional activity in the heart. About 1/3 of basal CREB/CBP interactions in heart cells depend on $Ca_V1.2$ channel activity but, unlike in neurons, are not affected by CaMKII inhibition. Despite the fact that major E-T coupling signaling molecules in all excitable cells are the same, the fine intranuclear architecture and dynamics of CREB-dependent transcriptional signaling might be different in cardiomyocytes and neurons.

It is worth reminding that all major components of $Ca_V1.2$ are subject to alternative splicing that generates a large diversity of its variants with poorly investigated distinctions in the functional properties. In addition, $Ca_V1.2$ channels tend to form large clusters in the plasma membrane. The size of these clusters and their number in the plasma membrane depend on the type of accessory β-subunits (Kobrinsky et al., 2009). Moreover, β-subunits add further complexity to the channel regulation by their tendency towards homo- and hetero-oligomerization (Lao et al., 2010). It remains unknown whether these properties affect CDT and are utilized in nature for fine tuning or pathogenic change of the CREB-dependent transcriptional response. It is known, however, that one of the α_{1C} splice variants is expressed predominantly in atherosclerotic vascular smooth muscle cells (Tiwari et al., 2006). Whether the association of this "atherosclerotic" splice variant with the pathogenic cell proliferation in atherosclerosis is mediated by a specific intranuclear architecture and the dynamics of CREB-dependent transcriptional signaling (Soldatov, 2013) is an important question for future research.

This study was supported by the National Institute on Aging Intramural Research Program (Z01 AG000294-08 to NMS).

References

Bregman, D. B., Du, L., van der Zee, S., and Warren, S. L. (1995). Transcription-dependent redistribution of the large subunit of RNA polymerase II to discrete nuclear domains. *J. Cell Biol.* 129, 287–298.

Bers, D. M. (2008) Calcium cycling and signaling in cardiac myocytes. *Annu. Rev. Physiol.* 70, 23–49.

Bers, D. M. (2011). Ca^{2+}-calmodulin-dependent protein kinase II regulation ofcardiac excitation-transcription coupling. *Heart Rhythm,* 8, 1101-1104.

Dolmetsch, R. E., Pajvani U., Fife, K., Spotts, J. M., Greenberg, M. E., (2001), Signaling to the nucleus by an L-type calcium channel-calmodulin complex through the MAP kinase pathway, *Science* 294, 333-339.

Domínguez-Rodríguez, A., Ruiz-Hurtado, G., Benitah, J.-P., and Gómez, A. M. (2012) The other side of cardiac Ca^{2+} signaling: transcriptional control. *Frontiers Physiol.,* 3, 1-7.

Flavell, S. W. and Greenberg, M. E. (2008) Signaling mechanisms linking neuronal activity to gene expression and plasticity of the nervous system. *Annu. Rev. Neurosci.* 31, 563–590.

Greer, P. L. and Greenberg, M. E. (2008) From synapse to nucleus: calcium-dependent gene transcription in the control of synapse development and function. *Neuron,* 846-860.

Grinsted, A., Moore, J. C., Jevrejeva, S. (2004). Application of the cross wavelet transform and wavelet coherence to geophysical time series, *Nonlinear Processes in Geophysics* 11, 561-566.

Halling, D. B., Aracena-Parks, P. and Hamilton, S. L. (2005) Regulation of voltage-gated Ca^{2+} channels by calmodulin. *Sci. STKE,* re15.

Klevecz, R. R. and Li, C. M. (2007) Evolution of the clock from yeast to man by period-doubling folds in the cellular oscillator. *Cold Spring Harbor Symp. Quant. Biol.* 72, 421–429.

Kobrinsky, E., Schwartz, E., Abernethy, D. R. Soldatov, N. M. (2003). Voltage-gated mobility of the Ca2+ channel cytoplasmic tails and its regulatory role, *J. Biol. Chem.* 278, 5021-5028.

Kobrinsky, E., Mager, D., Bentil, S., Murata, S., Abernethy, D. R., Soldatov, N. M. (2005), Identification of plasma membrane macro- and microdomains from wavelet analysis of FRET microscopy, *Biophys. J.* 88, 3625-3634.

Kobrinsky, E., Abrahimi, P., Duong, S. Q., Thomas, S., Harry, J. B., Patel, C., Lao, Q. Z., and Soldatov, N. M. (2009) Effect of $Ca_v\beta$ subunits on structural organization of $Ca_v1.2$ calcium channels. *PLoS ONE* 4, e5587.

Kobrinsky, E., Duong, S. Q., Sheydina, A., Soldatov, N. M. (2011) Microdomain organization and frequency-dependence of CREB-dependent transcriptional signaling in heart cells. *FASEB J.,* 25:1544-1555.

Koch, M., Mauhin, V., Stehle, J. H., Schomerus, C., and Korf, H.-W. (2003). Dephosphorylation of pCREB by protein serine/threonine phosphatases is involved in inactivation of *Aanat* gene transcription in rat pineal gland. *J. Neurochem.* 85, 170–179.

Kornhauser, J. M., Cowan, C. W., Shaywitz, A. J., Dolmetsch, R. E., Griffith, E. C., Hu, L. S., Haddad, C., Xia, Z., and Greenberg, M. E. (2002). CREB transcriptional activity in neurons is regulated by multiple, calcium-specific phosphorylation events. *Neuron* 34, 221–233.

Mager, D. E., Kobrinsky, E., Masoudieh, A., Maltsev, A., Abernethy, D. R., Soldatov, N. M. (2007). Analysis of functional signaling domains from fluorescence imaging and the two-dimensional continuous wavelet transform, *Biophys. J.*, 93, 2900-2910.

Maharana, S., Sharma, D., Shi, X., Shivashankar, G V. (2012). Dynamic organization of transcription compartments is dependent on functional nuclear architecture. *Biophys. J.,* 103, 851-859.

Özgen, N., Rosen, M. R. (2009) Cardiac Memory: work in progress. *Heart and Rhythm,* 564-570.

Parada, L. A., McQueen, P. G. and Misteli, T. (2004) Tissue-specific spatial organization of genomes. *Genome Biol.* 5, R44.

Patberg, K. W. and Rosen, M. R. (2003) On the role of the cAMP response element binding protein in long-term cardiac memory. *Circ. Res.* 93, e87.

Rosen, M. R., Cohen, I. S. (2006). Cardiac memory ... new insights into molecular mechanisms, *J. Physiol.* 570, 209-218.

Sands, W. A. and Palmer, T. M. (2008) Regulating gene transcription in response to cyclic AMP elevation, *Cell. Signalling*, 20, 460–466.

Satin, J., Itzhaki, I., Rapoport, S., Schroder, E. A., Izu, L., Arbel, G., Beyar, R., Balke, C. W., Schiller, J., and Gepstein, L. (2008) Calcium handling in human embryonic stem cell-derived cardiomyocytes. *Stem Cells* 26, 1961–1972.

Soldatov, N. M. (2003) Ca^{2+} channel moving tail: link between Ca^{2+}-induced inactivation and Ca^{2+} signal transduction. *Trends Pharm. Sci.* 24, 167-171.

Soldatov, N. M. (2012) Molecular determinants of $Ca_v1.2$ calcium channel inactivation. *ISRN Mol. Biol.* (Article ID 691341), 1-10.

Soldatov, N. M. (2013) $Ca_v1.2$, cell proliferation, and new target in atherosclerosis. *ISRN Biochemistry* (Article ID 463527), 1-13.

Tiwari, S., Zhang, Y., Heller, J., Abernethy, D. R., and Soldatov, N. M. (2006) Atherosclerosis-related molecular alteration of the human $Ca_v1.2$ calcium channel α_{1C} subunit. *Proc. Natl. Acad. Sci. US* 103, 17024-17029.

Trinkle-Mulcahy, L. and Lamond, A. I. (2008) Nuclear functions in space and time: Gene expression in a dynamic, constrained environment. *FEBS Lett.* 582, 1960–1970.

Wheeler, D. G., Barrett, C. F., Groth, R. D., Safa, P., Tsien, R. W. (2008), CaMKII locally encodes L-type channel activity to signal to nuclear CREB in excitation-transcription coupling, *J. Cell. Biol.* 183, 849-863.

Wheeler, D. G., Groth, R. D., Curtis, H. M., Barrett, C. F., Owen, S. F., Safa, P., Tsien, R. W. (2012), Ca_V1 and Ca_V2 channels engage distinct modes of Ca^{2+} signaling to control CREB-dependent gene expression, *Cell* 149, 1112-1124.

In: New Developments in Calcium … ISBN: 978-1-62948-601-7
Editor: Masayoshi Yamaguchi © 2014 Nova Science Publishers, Inc.

Chapter IV

A Novel Concept on the Repetitive Calcium Elevation

Ruei-Cheng Yang[1], Wen-Li Hsu[1] and Tohru Yoshioka[2]
[1]Department of Physiology, School of Medicine,
Kaohsiung Medical University, Kaohsiung, Taiwan
[2]Graduate Institute of Medicine, Kaohsiung Medical University,
Kaohsiung, Taiwan

Abstract

Not only calcium transient but also repetitive calcium elevation plays an important role in the elucidation of cellular responses for various types of stimulations. The pattern of calcium oscillation can be classified into three groups according to frequency; normal (less than 1 Hz), slower (milli Hz) and ultra-low (micro Hz). These three groups of calcium oscillation are associated with different values of temperature dependent coefficient, Q10. The Q10 value for normal frequency was found to be about 4, which corresponds to the ATP consumption process. The value of Q10 = 2 is well known for general biochemical reactions, while Q10 = 1 matches the physical process; ionic diffusion.

Normal frequency with Q10 = 4 is valid for keeping secretion constant for a long time. On the other hand, ordinary slower oscillation with Q10 = 2 means that the calcium oscillation is guaranteed by enzymatic reactions. The rate determinant process of ultra-low frequency oscillation is due to ionic diffusion (Q10 = 1.0), where the diffusion coefficient of ions is exceptionally reduced due to bound water on the surface of proteins. Although intracellular water structure is highly dependent on the cell cycle; bulk (free) water in G1 and G2, and bound water in the S and M phase. Since the period of G1 and G2 occupied 60 %~70% of the full period of one cell cycle, which means about 70% of cells in the in vitro experiment are staying at the G1 and G2 phase, on average. In the case of the circadian rhythm experiment, the cell cycle is arrested, thus the obstacle of ionic diffusion by bound water must be considered. Eventually a role of calcium oscillation may be a kind of cell clock in the cell cycle and circadian rhythm, while time interval calcium elevation is needed for the secretion process to avoid cell death.

Introduction

When a cell is stimulated with various types of signals, cells will respond to the stimulation by the elevation of intracellular calcium with a variety of modes. Except for the transient increase of intracellular calcium, many types of cells show continuous periodic oscillation of calcium levels for a single type of stimulation. According to Berridge [1], these oscillations can be explained by the two pool model, which was developed to explain how inositol 1, 4, 5-triphosphate (IP$_3$) set up primer Ca^{2+} elevation from internal stores via activation of IP$_3$ receptors (IP$_3$Rs) together with an influx of external Ca^{2+} [2]. This primer Ca^{2+} cannot elevate cytosolic calcium because IP3 insensitive ER rapidly sequesters it. Once the ER was filled with calcium, the stored calcium is released by calcium induced calcium release and makes a self-propagating wave.

At that time it was already known that the calcium oscillation appeared in various types of stimulation, such as neurotransmitters, hormones, growth factors, antigen and sperm. Recently, another model has been available because many other molecular operations were discovered such as transient receptor potential (TRP) channels and/or space operated channels (SOCs) other than voltage dependent calcium channels (VOCs). Among them both TRP and SOCs are closely connected to IP$_3$ induced calcium release, which make the oscillation mechanism more complex. In addition to those, several types of receptors and channels were registered in the list of functional proteins for calcium release, such as ryanodine receptor (RyR), and direct coupling between ER and mitochondria (MFN 1 and 2) and calcium ATPase [3].

What is the purpose of cytosolic calcium oscillation? Berridge prepared the answer to the question; a study of such oscillatory activity will help us to dissect and identify the major mechanism responsible for regulating intracellular calcium. Furthermore he described that the existence of calcium oscillations may be employing an inevitable consequence of employing influx as the primary mechanism for initiating a calcium signal. In this chapter, however, we will attempt to make another possibility of repetitive calcium oscillation of living cells based on the recent advances of cell biology.

Before considering a role of repetitive calcium elevation hitherto accepted, we will reconsider the general role of repetitive Ca^{2+} elevation in the living cell, because there is an absolutely large difference in characteristics between primary and cancer cells regarding cell cycle mode; cell cycle arrest in the primary cells and cell cycle release in the cancer cells. Calcium signaling in the fertilized egg, stem cell and iPS will belong to this category, a huge number of experiments have been performed using cell line cells (cancer cells) but the resulting models were used to explain the mechanism of primary cells in the body. Therefore we will initiate our discussion with different aspects from previously proposed models.

New Molecular Stuff on the Endoplasmic Reticulum (ER) Membrane Involved in Calcium Oscillations

For calcium release from ER, IP_3R, RyR and uptake of calcium by sarco/endoplasmic reticulum calcium transport ATPase (SERCA) were already established. Among them IP_3R has 3 subtypes: IP_3R1 can induce a transient and/or oscillatory calcium signal, IP_3R2 is the main oscillatory subtype and IP_3R3 shows anti-oscillatory behavior [4]. Also, the involvement of RyR for the repetitive calcium elevation is clear at present. The role of ryanodine sensitive calcium store in human oocytes was initially reported by Sousa et al that ryanodine induced calcium oscillation in the human oocytes at fertilization [5].

Opposite to that, however, spermatozoa or thimerosal induced calcium oscillation in mammalian oocyte was stopped by ryanodine. In order to solve this problem, the two store model, one is ryanodine sensitive the other is ryanodine insensitive, was proposed but it is still unclear [6]. In comparison with it, a role of SERCA (calcium ATPase) is more evident. Based on numerous data of calcium oscillation of Xenopus oocyte, which has no ryanodine receptor, several mathematical models for calcium oscillation were proposed without RyR. Falcke et al. found that over expression of SERCA lead to a decrease in the period (higher frequency) and an increase in the amplitude of the intracellular calcium wave [7]. Direct coupling between ER-SERCA and intracellular energy generation sites, mitochondria, was found to explain not only simple periodic oscillation but also bursting and spiking [8].

Significance of Calcium ATPase of ER for Maintaining Calcium Oscillations

Many of the functional proteins have been proposed for the explanation of initiation and maintenance of repetitive calcium elevation. A typical and the most popular crew forming IP_3 second messenger is the G protein coupled receptors (GPCR), trimetric G protein and phospholipase C (PLC). Entry of Ca^{2+} from extracellular space via SOC and ROC is necessary to fill the calcium stock of ER. The SOC channel, however, has a possibility to induce cytosolic calcium oscillation combined with IP_3R, SERCA and plasma membrane calcium ATPase (PMCA) [9].

The role of PMCA in spontaneous calcium oscillation of GH3 rat pituitary cells was further studied in detail by Hirono et al. and they found PMCA was not involved in calcium oscillation [10]. They suggested that in GH3 secretary cells, calcium influx via L-type calcium channel initiates spontaneous calcium oscillation, which are then maintained by the combined activity of calcium ATPase (SERCA) and calcium induced calcium release (CICR) from thapsigargin (Tg)–insensitive intracellular calcium stores. Furthermore they found amplitude of the spontaneous calcium oscillation and amount of calcium uptake into ER were temperature dependent linearly between 17°C and 28°C with Q10 = 4. Based on the discussions described above, the standard model of calcium oscillation mechanism is shown in Figure 1.

Utility of Parameter, Q10 Values, for Elucidating the Mechanism of Calcium Oscillation

As described above, calcium oscillation was found to be regulated largely dependent on temperature. In order to find out what type of molecular mechanism is involved in the oscillation mechanism, Q10 values is very valuable, because the usual biochemical reaction always obeys the rule of Q10 = 2. On the other hand, Barros and Kaczorowski demonstated that both Na^{+}-

and ATP- driven calcium uptake was reduced to some extent (10% ~ 20%) by lowering temperature from 37°C to 25°C in GH3 vesicles (Q10 = 1.1 ~ 1.2) [11].

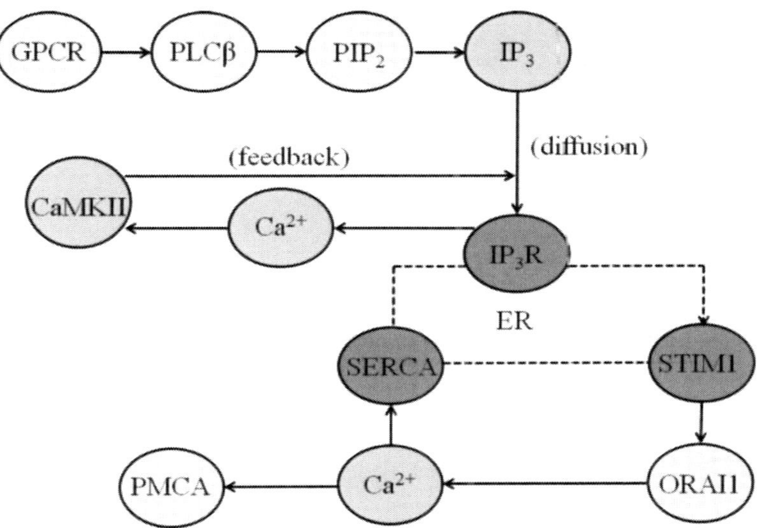

Figure 1. Schematic diagram of normal (~ Hz) and slower (milli Hz) calcium oscillation. In the normal range of calcium oscillation (Q10 = 4), SERCA and PMCA are rate determinant, while in the slower oscillation (Q10 = 2), the enzymatic reaction of PLC and CaMKII are rate determinant. In both cases, diffusion of IP$_3$ and Ca^{2+} were not disturbed by water structure, because the calcium experiment is possible only for the period of G1 or G2. In this figure, molecules of white color are in the plasma membrane; those of red color are on the ER membrane and blue ones in a cytosol. Abbreviation: GPCR, G protein coupled receptors; PLCβ, phospholipase C beta; PIP2, phosphatidylinositol 4, 5-bisphosphate; IP3, inositol 1, 4, 5-triphosphate; CaMKII, calcium/calmodulin-dependent protein kinase II; Ca^{2+}, calcium; IP3R, inositol 1, 4, 5-triphosphate receptor; SERCA, sarco/endoplasmic reticulum calcium transport ATPase; PMCA, plasma membrane calcium ATPase.

A similar low value of Q10 (= 1.6) for temperature range from 37°C ~ 17°C for Na$^+$/ Ca^{2+} exchange through sarcolemma has been reported in cultured cardiac myocyte [12]. The Q10 values around 1.0 suggest that the Na$^+$/ Ca^{2+} exchange process may be similar to the diffusion process not to the energy dependent one. The high temperature dependence Q10 = 4, however, is consistent with the idea of the energy dependent process, such as calcium uptake and/or extrusion by calcium ATPase, because the Q10 value of 2.5 ~

3.5 was found for the calcium pump [13-14] and a Q10 value of 3.6 ~ 4.0 for the Na^+/ Ca^{2+} exchanger [15-17], respectively. The strong temperature dependence of the oscillation frequency around Q10 = 3.0 for vasopressin activated mesangial cells, which is probably a mixture of the usual biochemical reaction and the ATP dependent process [18].

It is also likely that the transient receptor potential protein family C (TRPC) channel is involved in the formation of calcium oscillation. Recently participation of the TRPC3 channel in calcium oscillations is mediated by OX_1 orexin receptors with an activation of the TRPC3 channel [19]. Calcium oscillation in C. elegance intestinal epithelial cells was also found to be regulated by oscillatory activity of the transient receptor potential protein family M (TRPM) 7 (TRPM7) [20], and it was also established that stimulated orexine receptor generated IP_3 and diacylglycerol (DG), which can activate TRPCs [21].

Drosophila TRPCs (TRPC3, TRPC6 and TRPC7), however, are activated by DG, and two others, TRPC4 and TRPC5, are activated as a consequence of PLC activity [22]. This types of confusing can be avoided by using uncaged IP_3. Interestingly TRPM8, activated by cold temperature, voltage and menthol, has a very high Q10 value of 24 between 18°C and 25°C [23]. Also, Q10 (>10) was found for transient receptor potential protein family V (TRPV) 1 to 4 and other many types of TRP channels [24]. Therefore it is reasonable to presume that IP_3 and DG activated TRP channels must be involved in calcium signaling. If that is the case we have to pay attention to another possibility that type3 IP_3R has the possibility to be a kind of TRPC, which is activated by store depletion dependent calcium entry [25-26].

A Relationship between Frequency of Calcium Oscillation and Temperature Dependency (Q10 value) in the Living Cell (Tentative Hypothesis)

What is the role of calcium oscillation? From the moment of incidence to termination of life, calcium is the most important signaling molecule in living cells, because it has more than 20,000 times concentration gradient between intracellular and extracellular space. Different from single calcium transient response, a repetitive calcium response plays a more important role for cell cycle regulation, cell division, cell differentiation, cell proliferation and apoptosis. Not only in primary cells, but also in stem cells and cancer cells, calcium oscillation may be favorable for propagating important signals to whole cells and keep time like a clock in several cases. In order to elucidate the role of calcium oscillation more clearly, here we attempt to classify the oscillation mode into three groups depending on frequency; normal (< 1Hz), lower (~ milli Hz) and ultra-low (~micro Hz). In Figure 1, scheme to form calcium oscillation pattern is depicted for normal and slower frequency.

1) Normal frequency (< 1Hz) of calcium oscillation with Q10 = 4 is for secretion:

   The initial study of calcium oscillation started at the beginning of 1990 in connection with secretion mechanism using parotid acinar cells [27], rat chromaffin cells [28], rat pituitary cells [29], salivary acinar cells [30], parathyroid cells [31], hypothalamic neurons [32] and megakaryocyte [17]. In these studies various types of inhibitors were used for the identification of enzymes related to calcium oscillations, but the relative contribution of each component of cell signaling was not well characterized at that time. As was described previously, Arlenius plot analysis to evaluate Q10 values was found to be valuable to determine if the processes are obeyed in an energy dependent way or due to biochemical reaction or due to the physiological diffusion process. Based on these values, calcium oscillation with Q10~4 (calcium pump system) is reasonable to achieve long lasting continuous secretion, because the cell has to have a system to increase and reduce calcium levels quickly and continue for long periods without cell damage.

2) Lower frequency (~ mHz) of calcium oscillation with Q10 = 2 for cell cycle regulation.

   All of these lower frequencies of calcium oscillation were obtained when mammalian egg and/or Xenopus oocytes were studied to clarify the role of calcium oscillation for the cell cycle [33-34] and for egg to embryo transition [35]. In mouse embryonic stem cells (mES), the undifferentiated one exhibited different patterns of spontaneous calcium oscillations during cell cycle progression. It is also determined that the calcium oscillations were not initiated by Ca^{2+} influx but depend on IP_3 mediated Ca^{2+} release and the refilling of ER by SOC mechanism. Cell cycle analysis disclosed that IP_3 mediated Ca^{2+} release (8~12 mHz) is necessary for cell cycle progression through G(1)/S [36]. In addition to that, Hu and his group suggested that several mHz orders of intracellular calcium oscillation is valid for transcriptional activity using human umbilical vein endotherial

(HUVEC) cells, human bronchial epitherial (HBE) cells and human aortic endotherlial (HAE) cells [37]. Furthermore, about 10 mHz of calcium oscillation has the possibility to trigger nuclear factor activation, because Lewis suggested calcium oscillation (from 4 to 8mHz) may enhance both the efficiency and specificity of signaling through the calcium dependent transcription factors and the nuclear factor of activated T-cells [38].

Calcium signaling in eggs of many species, such as sea urchins and starfish at fertilization, ended by the explosion of calcium release arising from the latent period [39]. However, mammalian eggs and a small number of other organisms never stopped calcium signals at the first explosion but rather continued in the form of a series of calcium transients, which continued for 3~4 hours [40]. Why do mammalian eggs show such a lasting calcium oscillation? There must be two underlying reasons for the difference between the mammal and other species with regard to their reproduction system. According to Jones (1998), the reason is that mammalian fertilization is carried out internally and as a result of this, there is a marked delay between ovulation of eggs arrested at MII and fertilization [41]. The time at which the egg is actually arrested is dependent on the species, but the mechanism the sperm has adopted to relieve arrest (calcium release) is the same for all. Not only in cell division, calcium transients are always associated with the cleavage period and essential features of embryonic cytokines [42]. Recently, Lee et al. proposed a new model for cell cycle associated calcium oscillation in mammalian eggs; in the meta phase (MII), IP_3R1 undergoes degradation upon fertilization, then in interphase, IP_3R1, which is phosphorylated during MII stage, phosphorylation is nearly lost in interphase [43]. Cell cycle coupled calcium oscillation was discussed further by Jellerette et al, who showed that calcium oscillation and the cell cycle are entrained in the mouse egg and that IP_3R1 is phosphorylated in the cell cycle–dependent manner [34]. Temperature dependence of IP_3 involved calcium wave signaling in the Xenopus oocyte, which was observed by Leckleiter, and they obtained Q10 value as 1.5, which is less than the expected value (Q10 = 2). This lowered Q10 value may be due to the diffusion process for IP_3 [44]. In the Xennopus oocyte the cytosolic diffusion constant was estimated as 300 microns (+2)/second and that for Ca^{2+} was 20 microns (+2)/second [45] (Figure 1).

3) Ultra-low frequency calcium oscillation with Q10 = 1.0 of circadian rhythm.

The lowest frequency of calcium oscillation may be circadian frequency; one periodic change in a day (1 cycle/day= 0.000011 Hz = 11 microHz). Initially we attempted to observe change in the intracellular calcium level of SCN neurons using fura-2 fluorescent dye, which is known as calcium chelator. In this experiment the calcium level of SCN was higher in the daytime and lower in the nighttime [46]. The obstacle of long period calcium measurement was overcome by introducing genes of the calcium sensitive florescent protein, cameleon, in the living cell [47]. This technique was applied to SCN neurons and for circadian cytosolic calcium oscillation successfully [48]. Since this oscillation was found to be ryanodine sensitive, it was once concluded the circadian Ca^{2+} oscillation could be due to release of Ca^{2+} from ryanodine sensitive store. These findings were supported later by Hong et al, also using cameleon, and they additionally found that circadian waves of cytosolic Ca^{2+} was important for long range network connection of SCN in achieving its phase and period coherence [49].

What type of circadian calcium oscillation mechanism is likely including RyR? Theoretically, it was established that rhythmic oscillation could be produced by positive feed-back loop in the cell [50]. Essentially this loop must have molecular components, cGMP, cADPR and RyR or cAMP, IP_3 and IP_3R. According to O'Neil, molecular models for mammalian circadian rhythm formation have focused largely on transcriptional-translational feedback loops. And they made a new model, which determines the canonical pacemaker properties of amplitude, phase and frequency [51]. Also circadian calcium oscillation of rat SCN was confirmed by using Ca^{2+} sensitive cameleon [49]. Separately, another model that the GABAergic network in SCN is a key regulator of the biological clock was proposed by Ganguly and Lee [52]. These findings, along with the line GABA as the main neurotransmitter of SCN neuron, may play a key role in the function of the master circadian pacemaker [52]. Although cAMP dependent PKA could modulate topiramate potentiation of the GABA (A) receptor, GABA will be more effective in the brain outside SCN [53]. However, positive feedback loop by cGMP, cADP ribose (cADPR), RyR and calcium is more likely, although it was proposed for mHz of slower calcium oscillation [54].

It was first discovered that cADPR could activate RyR in 1991. In 1989 cADPR was identified as the active metabolite of nicotinamide adenine dinucleotide (NAD) [55]. Twenty years or more later, CD38 was found to be a multi-functional enzyme that can generate cADPR by consuming NAD^+ [56]. As Turovsky et al proposed, novel positive feed back loop by NO, cGMP, cADPR and Ca^{2+} will be applicable to circadian oscillation of SCN as it is except ionic diffusion constant. Recently, RyR was found to be the leading character in the circadian calcium rhythm formation with positive feed-back, because RyR was regulated by the circadian clock in reverse [57]. Furthermore, cADPR was synthesized by cGMP dependently [58], and melatonin, which is known as master clock output and internal time giver, modulates cGMP signaling [59]. The possibility of coupling cADPR with cAMP, which was discovered by Xie in cadiomyocyte [60], could be excluded although cAMP and Ca^{2+} oscillation was found to link IP_3-calcium signaling [61], because PLC beta knock out mouse lacking IP_3 showed normal circadian rhythm [62].

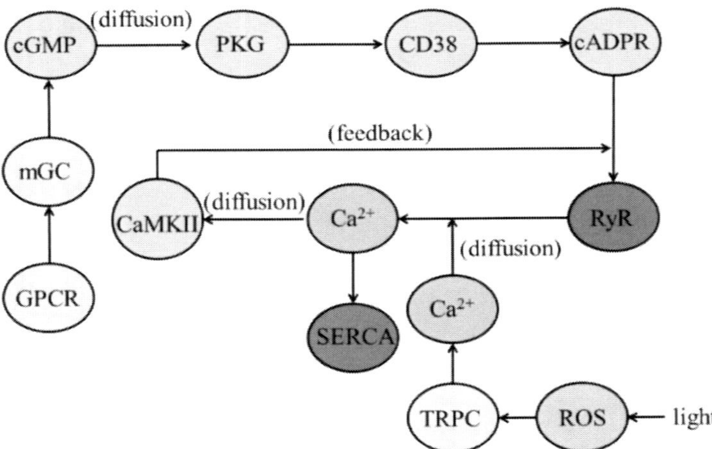

Figure 2. Schematic diagram of ultra slow (micro Hz) calcium oscillation. Circadian calcium oscillation 1 cycle/day) is typical. All primary cells in the body have no cell cycles (cell cycle arrest), where water in the cell is in a bound state, which largely obstructs diffusion of Ca^{2+} and cGMP. Since rate determinant in this case is the diffusion process (Q10 = 1), the disturbance of ionic diffusion process makes such a ultra low frequency. Abbreviation: mGC, membrane guanylyl cyclase; cGMP, cyclic guanosine monophosphate; PKG, protein kinase G; cADPR, cyclic adenosine diphosphoribose; RyR, ryanodine receptor; TRPC, transient receptor potential protein family C; ROS, reactive oxygen species.

Regrettably, there is no data of Q10 value for the circadian experiment, but the introduction of ultra slow diffusion of cGMP and Ca^{2+} in the primary cell has the possibility to explain the mechanism. As shown in Figure 2, the molecular mechanism for ultra-slow calcium oscillation (micro Hz) is likely if diffusion of cGMP and Ca^{2+} are significantly blocked by bound water on the protein surface. If that is the case Q10 could be estimated as 1.0. This value of Q10 = 1.0 agreed well with only one observed value Q10 = 1.05 for temperature range of 22 degree to 29 degree using "Euglena" [63].

The proposed blocking mechanism of ionic diffusion in the primary cell is discussed in the following section.

Effect of Cytosolic Water Structure on Ionic Diffusion Process

As was described above, there are two models for calcium oscillation; one is IP_3- IP_3R model for lower frequency (mili Hz) of fertilized egg and the other is cADPR- RyR model for ultra low frequency (micro Hz) of circadian rhythm. Both types of freoscillation models, however, were proposed initially for the secretion process using cell line (cancer) cells. Theoretically, as was shown in the Figures 1 and 2, calcium concentration can oscillate with normal frequency (< 1Hz) if second messengers (Ca^{2+} and cGMP) can move freely with the diffusion constant in normal (bulk) water. Actually, it is known that water structure in the egg is changed with the cell cycle. Furthermore, cytosol of the primary cell is considered as high density gel in the S phase and bulk water enriched sol in G1 and G2 phase (will be discussed in detail later), hence the mechanism for slowing down the calcium oscillation frequency must be discussed first and then make a novel model for slower and ultra-slow frequency of calcium oscillations.

Image for Cytosolic Water Structure

So far almost all biologists and biochemists imaged that water in the cell, which occupied 70 ~ 80 % of the cell mass, is abundant enough to allow diffusion of ions, second messengers and water-soluble proteins as in the test tubes. Physiologists, on the other hand, had full of realization that cytosol was a gel itself by the extrusion of endoplasm from squid giant axons for the voltage clamp experiment in 1970's. The long-lived ignorance for cytosolic water structure was also promoted by misinterpretation of the electron microscopic imaging [64]. Advances of physico-chemistry changed the image of intracellular water and it was established that two kinds of water existed in the cell; bulk (normal) water and bound (ordered structured) water. Eventually as was shown by Mentre, there is little space between protein molecules, which constitute mechanical obstacles for ionic diffusion (Figure 3). Furthermore, water molecules are strongly attracted by ionized domain such as COO^- and NH^+ and form an electro-strict water molecule with a density higher than 1.0 [65]. When water molecules are bound to consecutive polar amino acids (Asn, Cys, Gln, Ser, Thr, Tyr) in a polypeptide chain, they can bind together to form H-bonded lines.

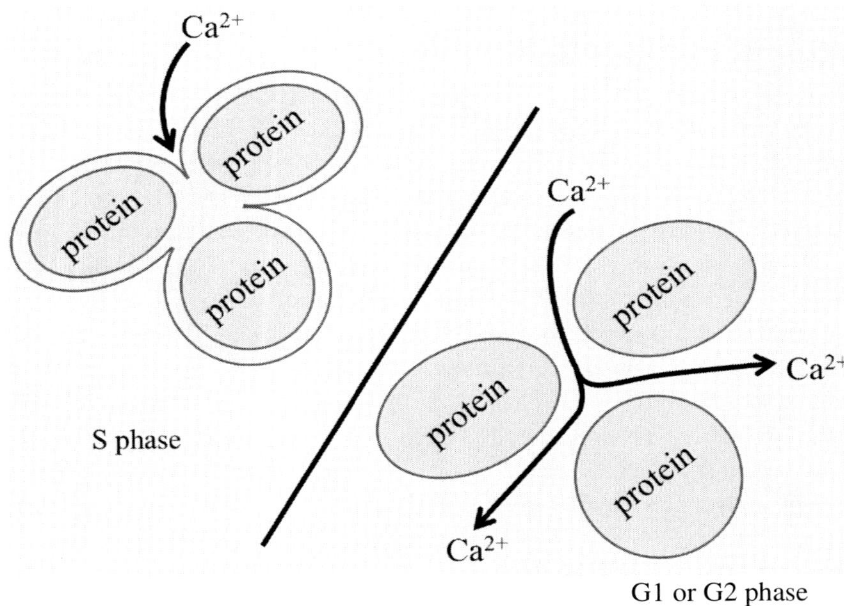

Figure 3. Schematic diagram for Ca^{2+} diffusion in cytosol at G1/G2 and S phase. In S phase surface of all proteins are covered by bound water, where no gap between proteins block ionic diffusion (upper part), while in G1 and G2 phase, bound water is changed to free (bulk) water where there is no obstruction for ionic diffusion. When cell cycle is released, 60 ~ 70 % of cells always stayed in G1 and/or G2 phase if cell cycles of each cell is randomized.

These water molecular lines are good conductor of protons [64]. When a proton arrives on the first water molecule of the water chain, the electrostatic equilibrium of a first water molecule of the chain is broken by the formation of OH3+, and then liberation of another proton will be induced. Just like domino toppling, this type of proton jump is continued to the last water molecule of this chain. This signal transfer can stimulate, for example, a calcium wave that can propagate along the protein surface [66].

The significance of proton flow in the cell signaling was proved by the water/heavy water (H/D) exchange; ion channel conductivity, calcium signaling and the cytoskeletal structure was found to be largely changed by the H/D exchange effect [67-69]. When H is replaced by D, the rate of proton liberation will be disturbed largely, because D has same charge (+1) with 2 times heavier mass than H.

Cell Cycle Is Another Key Word to Understand the Difference between the Primary Cell and Cancer (or Fertilized Egg) Cells

The Cancer cell is known to repeat cell division dragged for a long period (ageless) and fertilized eggs also repeat the cell cycle with cell division until embryo. Both of them have cell cycles, G1/ S / G2 / M (meta)/ M (ana). Unfortunately, primary cells obtained from human body for in vitro experiment is limited to neuron and oocyte, their cycle is arrested. On the other hand, the cell cycle of HeLa cell is associated with the change of intracellular water structure; NMR relaxation time (T1) of the intracellular proton was maximum in meta phase and minimum in S phase [70]. Subsequently, it was found that the established cell line maintained relaxation time similar to tumor tissue values [71-72]. Characterization of water in unfertilized (cell cycle arrest) and fertilized (cell cycle release) sea urchin egg was studied with the fast proton diffusion (FPD) method and they found that fertilization induced the formation of bulk (normal) water from bound water [73]. Also relaxation time T1 is highly associated with cell cycle change [74-76]. The water structure change associated with cell cycle was thus confirmed.

Role of Intracellular Calcium Oscillation as a Concluding Remark

In this chapter, a wide range frequency (from Hz to micro Hz) of calcium oscillation in the living cell was demonstrated. The mechanism of calcium oscillation was described successfully by a single pool model with calcium ATPase, and these 3 types of frequency can be explained by slowing the feedback rate via reduction of the driving force for 2nd messengers in the cell.

The values of Q10 are available for mechanism; Q10 = 4 for ATP consumption process, Q10= 2 for general biochemical reaction and Q10 = 1 is for the ionic diffusion process, although it is highly dependent on the state of gel in the cytosol.

However it is still questionable what the role of calcium oscillation is in the cell.

As was predicted by Parkash and Kamlesh, there are many candidates of calcium oscillation associated phenomena such as transcription, cell cycle, differentiation, angiogenesis, teromerase, motility and apoptosis [77]. Different from cell cycles in cancer cells, mHz of calcium oscillation may be the cellular clock for activation of the egg and development of the embryo with sequential biochemical reaction, while circadian calcium oscillation is obviously a sandglass for shortening the telomere to determine life span of living things.

References

[1]   Berridge, M. J. Cytoplasmic calcium oscillations: a two pool model. *Cell calcium,* 1991 12, 63-72.

[2]   Miyazaki, S.; Yuzaki, M.; Nakada, K.; Shirakawa, H.; Nakanishi, S.; Nakade, S.; Mikoshiba, K. Block of Ca^{2+} wave and Ca^{2+} oscillation by antibody to the inositol 1,4,5-trisphosphate receptor in fertilized hamster eggs. *Science,* 1992 257, 251-255.

[3]   Decuypere, J. P.; Monaco, G.; Missiaen, L.; De Smedt, H.; Parys, J. B.; Bultynck, G. IP(3) Receptors, Mitochondria, and Ca Signaling: Implications for Aging. *Journal of aging research,* 2011 2011, 920178.

[4]   Zhang, S.; Fritz, N.; Ibarra, C.; Uhlen, P. Inositol 1,4,5-trisphosphate receptor subtype-specific regulation of calcium oscillations. *Neurochemical research,* 2011 36, 1175-1185.

[5]   Sousa, M.; Barros, A.; Tesarik, J. The role of ryanodine-sensitive Ca^{2+} stores in the Ca^{2+} oscillation machine of human oocytes. *Molecular human reproduction,* 1996 2, 265-272.

[6]   Tesarik, J.; Sousa, M. Mechanism of calcium oscillations in human oocytes: a two-store model. *Molecular human reproduction,* 1996 2, 383-386.

[7] Falcke, M.; Li, Y.; Lechleiter, J. D.; Camacho, P. Modeling the dependence of the period of intracellular Ca^{2+} waves on SERCA expression. *Biophysical journal*, 2003 85, 1474-1481.

[8] Dellen, B. K.; Barber, M. J.; Ristig, M. L.; Hescheler, J.; Sauer, H.; Wartenberg, M. [Ca^{2+}] oscillations in a model of energy-dependent Ca^{2+} uptake by the endoplasmic reticulum. *Journal of theoretical biology*, 2005 237, 279-290.

[9] Parekh, A. B. Decoding cytosolic Ca^{2+} oscillations. *Trends in biochemical sciences*, 2011 36, 78-87.

[10] Hirono, M.; Takamura, K.; Ito, Y.; Nakano, Y.; Chikaoka, Y.; Suzuki, N.; Yoshioka, T. Role of Ca(2+)-ATPase in spontaneous oscillations of cytosolic free Ca^{2+} in GH3 rat pituitary cells. *Cell calcium*, 1999 25, 125-135.

[11] Barros, F.; Kaczorowski, G. J. Mechanisms of Ca^{2+} transport in plasma membrane vesicles prepared from cultured pituitary cells. II. (Ca^{2+} + Mg^{2+})-ATPase-dependent Ca^{2+} transport activity. *The Journal of biological chemistry*, 1984 259, 9404-9410.

[12] Marengo, F. D.; Wang, S. Y.; Langer, G. A. The effects of temperature upon calcium exchange in intact cultured cardiac myocytes. *Cell calcium*, 1997 21, 263-273.

[13] Stein, R. B.; Gordon, T.; Shriver, J. Temperature dependence of mammalian muscle contractions and ATPase activities. *Biophysical journal*, 1982 40, 97-107.

[14] Sunano, S.; Moriyama, K.; Shimamura, K. The relaxation and Ca-pump inhibition in mesenteric artery of normotensive and stroke-prone spontaneously hypertensive rats. *Microvascular research*, 1991 42, 117-124.

[15] Kimura, J.; Miyamae, S.; Noma, A. Identification of sodium-calcium exchange current in single ventricular cells of guinea-pig. *The Journal of physiology*, 1987 384, 199-222.

[16] Russell, J. M.; Blaustein, M. P. Calcium efflux from barnacle muscle fibers. Dependence on external cations. *The Journal of general physiology*, 1974 63, 144-167.

[17] Tertyshnikova, S.; Fein, A. [Ca^{2+}]i oscillations and [Ca^{2+}]i waves in rat megakaryocytes. *Cell calcium*, 1997 21, 331-344.

[18] Hajjar, R. J.; Bonventre, J. V. Oscillations of intracellular calcium induced by vasopressin in individual fura-2-loaded mesangial cells. Frequency dependence on basal calcium concentration, agonist

concentration, and temperature. *The Journal of biological chemistry,* 1991 266, 21589-21594.

[19] Peltonen, H. M.; Magga, J. M.; Bart, G.; Turunen, P. M.; Antikainen, M. S.; Kukkonen, J. P.; Akerman, K. E. Involvement of TRPC3 channels in calcium oscillations mediated by OX(1) orexin receptors. *Biochemical and biophysical research communications,* 2009 385, 408-412.

[20] Estevez, A. Y.; Strange, K. Calcium feedback mechanisms regulate oscillatory activity of a TRP-like Ca^{2+} conductance in C. elegans intestinal cells. *The Journal of physiology,* 2005 567, 239-251.

[21] Ekholm, M. E.; Johansson, L.; Kukkonen, J. P. Rapid and easy semi-quantitative evaluation method for diacylglycerol and inositol-1,4,5-trisphosphate generation in orexin receptor signalling. *Acta physiologica,* 2010 198, 387-392.

[22] Putney, J. W., Jr. Inositol lipids and TRPC channel activation. *Biochemical Society symposium,* 2007, 37-45.

[23] Brauchi, S.; Orio, P.; Latorre, R. Clues to understanding cold sensation: thermodynamics and electrophysiological analysis of the cold receptor TRPM8. *Proceedings of the National Academy of Sciences of the United States of America,* 2004 101, 15494-15499.

[24] Latorre, R.; Brauchi, S.; Orta, G.; Zaelzer, C.; Vargas, G. ThermoTRP channels as modular proteins with allosteric gating. *Cell calcium,* 2007 42, 427-438.

[25] Putney, J. W., Jr. Type 3 inositol 1,4,5-trisphosphate receptor and capacitative calcium entry. *Cell calcium,* 1997 21, 257-261.

[26] Boulay, G.; Brown, D. M.; Qin, N.; Jiang, M.; Dietrich, A.; Zhu, M. X.; Chen, Z.; Birnbaumer, M.; Mikoshiba, K.; Birnbaumer, L. Modulation of Ca(2+) entry by polypeptides of the inositol 1,4, 5-trisphosphate receptor (IP3R) that bind transient receptor potential (TRP): evidence for roles of TRP and IP3R in store depletion-activated Ca(2+) entry. *Proceedings of the National Academy of Sciences of the United States of America,* 1999 96, 14955-14960.

[27] Foskett, J. K.; Roifman, C. M.; Wong, D. Activation of calcium oscillations by thapsigargin in parotid acinar cells. *The Journal of biological chemistry,* 1991 266, 2778-2782.

[28] D'Andrea, P.; Zacchetti, D.; Meldolesi, J.; Grohovaz, F. Mechanism of $[Ca^{2+}]i$ oscillations in rat chromaffin cells. Complex Ca(2+)-dependent regulation of a ryanodine-insensitive oscillator. *The Journal of biological chemistry,* 1993 268, 15213-15220.

[29] Wagner, K. A.; Yacono, P. W.; Golan, D. E.; Tashjian, A. H., Jr. Mechanism of spontaneous intracellular calcium fluctuations in single GH4C1 rat pituitary cells. *The Biochemical journal,* 1993 292 (Pt 1), 175-182.

[30] Foskett, J. K.; Wong, D. C. [Ca^{2+}]i inhibition of Ca^{2+} release-activated Ca^{2+} influx underlies agonist- and thapsigargin-induced [Ca^{2+}]i oscillations in salivary acinar cells. *The Journal of biological chemistry,* 1994 269, 31525-31532.

[31] Ridefelt, P.; Bjorklund, E.; Akerstrom, G.; Olsson, M. J.; Rastad, J.; Gylfe, E. Ca(2+)-induced Ca^{2+} oscillations in parathyroid cells. *Biochemical and biophysical research communications,* 1995 215, 903-909.

[32] Charles, A. C.; Hales, T. G. Mechanisms of spontaneous calcium oscillations and action potentials in immortalized hypothalamic (GT1-7) neurons. *Journal of neurophysiology,* 1995 73, 56-64.

[33] Kline, D.; Kline, J. T. Repetitive calcium transients and the role of calcium in exocytosis and cell cycle activation in the mouse egg. *Developmental biology,* 1992 149, 80-89.

[34] Jellerette, T.; Kurokawa, M.; Lee, B.; Malcuit, C.; Yoon, S. Y.; Smyth, J.; Vermassen, E.; De Smedt, H.; Parys, J. B.; Fissore, R. A. Cell cycle-coupled [Ca($^{2+}$)](i) oscillations in mouse zygotes and function of the inositol 1,4,5-trisphosphate receptor-1. *Developmental biology,* 2004 274, 94-109.

[35] Ducibella, T.; Huneau, D.; Angelichio, E.; Xu, Z.; Schultz, R. M.; Kopf, G. S.; Fissore, R.; Madoux, S.; Ozil, J. P. Egg-to-embryo transition is driven by differential responses to Ca(2+) oscillation number. *Developmental biology,* 2002 250, 280-291.

[36] Kapur, N.; Mignery, G. A.; Banach, K. Cell cycle-dependent calcium oscillations in mouse embryonic stem cells. *American journal of physiology. Cell physiology,* 2007 292, C1510-1518.

[37] Hu, Q.; Deshpande, S.; Irani, K.; Ziegelstein, R. C. [Ca(2+)](i) oscillation frequency regulates agonist-stimulated NF-kappaB transcriptional activity. *The Journal of biological chemistry,* 1999 274, 33995-33998.

[38] Lewis, R. S. Calcium oscillations in T-cells: mechanisms and consequences for gene expression. *Biochemical Society transactions,* 2003 31, 925-929.

[39] Stricker, S. A. Comparative biology of calcium signaling during fertilization and egg activation in animals. *Developmental biology*, 1999 211, 157-176.

[40] Carroll, J. The initiation and regulation of Ca^{2+} signalling at fertilization in mammals. *Seminars in cell and developmental biology*, 2001 12, 37-43.

[41] Jones, K. T. Ca^{2+} oscillations in the activation of the egg and development of the embryo in mammals. *The International journal of developmental biology*, 1998 42, 1-10.

[42] Webb, S. E.; Li, W. M.; Miller, A. L. Calcium signalling during the cleavage period of zebrafish development. Philosophical transactions of the Royal Society of London. Series B, *Biological sciences*, 2008 363, 1363-1369.

[43] Lee, B.; Yoon, S. Y.; Fissore, R. A. Regulation of fertilization-initiated $[Ca^{2+}]i$ oscillations in mammalian eggs: a multi-pronged approach. *Seminars in cell and developmental biology*, 2006 17, 274-284.

[44] Lechleiter, J. D.; Clapham, D. E. Molecular mechanisms of intracellular calcium excitability in X. laevis oocytes. *Cell*, 1992 69, 283-294.

[45] Atri, A.; Amundson, J.; Clapham, D.; Sneyd, J. A single-pool model for intracellular calcium oscillations and waves in the Xenopus laevis oocyte. *Biophysical journal*, 1993 65, 1727-1739.

[46] Colwell, C. S. Circadian modulation of calcium levels in cells in the suprachiasmatic nucleus. *The European journal of neuroscience*, 2000 12, 571-576.

[47] Miyawaki, A.; Llopis, J.; Heim, R.; McCaffery, J. M.; Adams, J. A.; Ikura, M.; Tsien, R. Y. Fluorescent indicators for Ca^{2+} based on green fluorescent proteins and calmodulin. Nature, 1997 388, 882-887.

[48] Ikeda, M.; Sugiyama, T.; Wallace, C. S.; Gompf, H. S.; Yoshioka, T.; Miyawaki, A.; Allen, C. N. Circadian dynamics of cytosolic and nuclear Ca^{2+} in single suprachiasmatic nucleus neurons. *Neuron*, 2003 38, 253-263.

[49] Hong, J. H.; Jeong, B.; Min, C. H.; Lee, K. J. Circadian waves of cytosolic calcium concentration and long-range network connections in rat suprachiasmatic nucleus. *The European journal of neuroscience*, 2012 35, 1417-1425.

[50] Chen, X. F.; Li, C. X.; Wang, P. Y.; Li, M.; Wang, W. C. Dynamic simulation of the effect of calcium-release activated calcium channel on cytoplasmic Ca^{2+} oscillation. *Biophysical chemistry*, 2008 136, 87-95.

[51] O'Neill, J. S.; Maywood, E. S.; Hastings, M. H. Cellular mechanisms of circadian pacemaking: beyond transcriptional loops. *Handbook of experimental pharmacology*, 2013 217, 67-103.

[52] Ganguly, A.; Lee, D. Suppression of inhibitory GABAergic transmission by cAMP signaling pathway: alterations in learning and memory mutants. *The European journal of neuroscience*, 2013 37, 1383-1393.

[53] Simeone, T. A.; Wilcox, K. S.; White, H. S. cAMP-dependent protein kinase A activity modulates topiramate potentiation of GABA(A) receptors. *Epilepsy research*, 2011 96, 176-179.

[54] Turovsky, E. A.; Turovskaya, M. V.; Dolgacheva, L. P.; Zinchenko, V. P.; Dynnik, V. V. Acetylcholine Promotes Ca(2+)and NO-Oscillations in Adipocytes Implicating Ca(2+)-->NO-->cGMP-->cADP-ribose-->Ca(2+) Positive Feedback Loop - Modulatory Effects of Norepinephrine and Atrial Natriuretic Peptide. *PloS one*, 2013 8, e63483.

[55] Lee, H. C.; Walseth, T. F.; Bratt, G. T.; Hayes, R. N.; Clapper, D. L. Structural determination of a cyclic metabolite of NAD+ with intracellular Ca^{2+}-mobilizing activity. *The Journal of biological chemistry*, 1989 264, 1608-1615.

[56] Graeff, R.; Liu, Q.; Kriksunov, I. A.; Kotaka, M.; Oppenheimer, N.; Hao, Q.; Lee, H. C. Mechanism of cyclizing NAD to cyclic ADP-ribose by ADP-ribosyl cyclase and CD38. *The Journal of biological chemistry*, 2009 284, 27629-27636.

[57] Gamble, K. L.; Ciarleglio, C. M. Ryanodine receptors are regulated by the circadian clock and implicated in gating photic entrainment. *The Journal of neuroscience: the official journal of the Society for Neuroscience*, 2009 29, 11717-11719.

[58] Graeff, R. M.; Franco, L.; De Flora, A.; Lee, H. C. Cyclic GMP-dependent and -independent effects on the synthesis of the calcium messengers cyclic ADP-ribose and nicotinic acid adenine dinucleotide phosphate. *The Journal of biological chemistry*, 1998 273, 118-125.

[59] Stumpf, I.; Bazwinsky, I.; Peschke, E. Modulation of the cGMP signaling pathway by melatonin in pancreatic beta-cells. *Journal of pineal research*, 2009 46, 140-147.

[60] Xie, G. H.; Rah, S. Y.; Kim, S. J.; Nam, T. S.; Ha, K. C.; Chae, S. W.; Im, M. J.; Kim, U. H. ADP-ribosyl cyclase couples to cyclic AMP signaling in the cardiomyocytes. *Biochemical and biophysical research communications*, 2005 330, 1290-1298.

[61] Ikeda, M.; Nelson, C. S.; Shinagawa, H.; Shinoe, T.; Sugiyama, T.; Allen, C. N.; Grandy, D. K.; Yoshioka, T. Cyclic AMP regulates the calcium transients released from IP(3)-sensitive stores by activation of rat kappa-opioid receptors expressed in CHO cells. *Cell calcium*, 2001 29, 39-48.

[62] Ikeda, M.; Hirono, M.; Sugiyama, T.; Moriya, T.; Ikeda-Sagara, M.; Eguchi, N.; Urade, Y.; Yoshioka, T. Phospholipase C-beta4 is essential for the progression of the normal sleep sequence and ultradian body temperature rhythms in mice. *PloS one*, 2009 4, e7737.

[63] Anderson, R. W.; Laval-Martin, D. L.; Edmunds, L. N., Jr. Cell cycle oscillators. Temperature compensation of the circadian rhythm of cell division in Euglena. *Experimental cell research*, 1985 157, 144-158.

[64] Mentre, P. Water in the orchestration of the cell machinery. Some misunderstandings: a short review. *Journal of biological physics*, 2012 38, 13-26.

[65] Mentre, P.; Hui Bon Hoa, G. Effects of high hydrostatic pressures on living cells: a consequence of the properties of macromolecules and macromolecule-associated water. *International review of cytology*, 2001 201, 1-84.

[66] Teissie, J.; Prats, M.; Soucaille, P.; Tocanne, J. F. Evidence for conduction of protons along the interface between water and a polar lipid monolayer. *Proceedings of the National Academy of Sciences of the United States of America*, 1985 82, 3217-3221.

[67] Ikeda, M.; Suzuki, S.; Kishio, M.; Hirono, M.; Sugiyama, T.; Matsuura, J.; Suzuki, T.; Sota, T.; Allen, C. N.; Konishi, S.; Yoshioka, T. Hydrogen-deuterium exchange effects on beta-endorphin release from AtT20 murine pituitary tumor cells. *Biophysical journal*, 2004 86, 565-575.

[68] Hirakura, Y.; Sugiyama, T.; Takeda, M.; Ikeda, M.; Yoshioka, T. Deuteration as a tool in investigating the role of protons in cell signaling. *Biochimica et biophysica acta*, 2011 1810, 218-225.

[69] Grdadolnik, J.; Marechal, Y. Hydrogen-deuterium exchange in bovine serum albumin protein monitored by fourier transform infrared spectroscopy, part I: structural studies. *Applied spectroscopy*, 2005 59, 1347-1356.

[70] Beall, P. T.; Hazlewood, C. F.; Rao, P. N. Nuclear magnetic resonance patterns of intracellular water as a function of HeLa cell cycle. *Science*, 1976 192, 904-907.

[71] Beall, P. T.; Hazlewood, C. F.; Rutzky, L. P. NMR relaxation times of water protons in human colon cancer cell lines and clones. *Cancer biochemistry biophysics,* 1982 6, 7-12.

[72] Beall, P. T.; Narayana, P. A.; Amtey, S. R.; Spiga, L.; Intra, E.; Ridella, S.; Mela, G. S. The systemic effect of cancers on human sera proton NMR relaxation times. *Magnetic resonance imaging,* 1984 2, 83-87.

[73] Merta, P. J.; Fullerton, G. D.; Cameron, I. L. Characterization of water in unfertilized and fertilized sea urchin eggs. *Journal of cellular physiology,* 1986 127, 439-447.

[74] Cameron, I. L.; Cook, K. R.; Edwards, D.; Fullerton, G. D.; Schatten, G.; Schatten, H.; Zimmerman, A. M.; Zimmerman, S. Cell cycle changes in water properties in sea urchin eggs. *Journal of cellular physiology,* 1987 133, 14-24.

[75] Zimmerman, S.; Zimmerman, A. M.; Cameron, I. L.; Fullerton, G. D.; Schatten, H.; Schatten, G. Effects of cytoskeletal inhibitors on water proton relaxation time changes in unfertilized and fertilized sea urchin eggs. *Cell biology international reports,* 1987 11, 605-614.

[76] Zimmerman, S.; Zimmerman, A. M.; Fullerton, G. D.; Luduena, R. F.; Cameron, I. L. Water ordering during the cell cycle: nuclear magnetic resonance studies of the sea-urchin egg. *Journal of cell science,* 1985 79, 247-257.

[77] Parkash, J.; Asotra, K. Calcium wave signaling in cancer cells. *Life sciences,* 2010 87, 587-595.

Reviewed by Robin Irvine,
Department of Pharmacology,
School of the Biological Sciences,
University of Cambridge; Katsuhiko Mikoshiba,
Riken Brain Science Institute.

In: New Developments in Calcium …
Editor: Masayoshi Yamaguchi

ISBN: 978-1-62948-601-7
© 2014 Nova Science Publishers, Inc.

Chapter V

Calcium Binding Proteins in the Liver Fluke, *FASCIOLA HEPATICA*

*Sean L. Russell and David J. Timson**
School of Biological Sciences, Queen's University Belfast,
Medical Biology Centre, Belfast, UK

Abstract

The common liver fluke, *Fasciola hepatica*, is a parasite of mammals. In the western world its effects are largely felt on agriculture where infection of cows, sheep and other farm animals is estimated to cause millions of dollars of financial losses. In the developing world, the problem is even more serious with an estimated 7 million infected people and many millions more at risk of infection. Calcium signalling is of key importance in all eukaryotic species and recent discoveries of novel types of calcium binding proteins in liver flukes (and related trematodes) suggest that there may be calcium signalling processes which are unique to this group of organisms. If so, these pathways may provide potential targets for the design of novel anthelmintic drugs. Here, we review three main groups of *F. hepatica* calcium binding proteins: the FH8 family, the calmodulin family (FhCaM1, FhCaM2 and FhCaM3) and the EF-hand/dynein light chain family (FH22, FhCaBP3, FhCaBP4).

* Corresponding author: Telephone: +44(0)28 9097 5875, Fax:+44(0)28 9097 5877, Email: d.timson@qub.ac.uk

Considerable information has been gathered on the sequences, predicted structures and biochemical properties of these molecules. The challenge now is to understand their functions in the organism.

Introduction: *FASCIOLA HEPATICA* and its Significance

The common liver fluke, *Fasciola hepatica,* is a parasite responsible for fascioliasis which is mainly of agricultural concern in the developed world in that it predominately affects livestock such as cattle and sheep although it can also infect humans. *F. hepatica* can be found globally but in Asia and Africa the closely related species *Fasciola gigantica* is more common [1]. The pathology of *F. hepatica* infections caused is by the migration of newly excysted juveniles from the intestinal tract to the liver of mammals. This can lead to anaemia and hypoalbumanaemia which cause lead to death in sheep and calves. A secondary infection by the bacterium *Clostridium oedematiens* in the necrotic lesions caused by the migration of the juveniles to the liver can also occur in sheep. The economic loss associated with infection is caused by the reduction in meat and milk yield in cattle and loss in wool growth and meat production in sheep [2].

Human fascioliasis is of emerging concern, with reports suggesting an increase in prevalence throughout the world [3]. Studies suggest that areas of relatively high human infection are not necessarily areas where livestock infection is high. This is probably due to differences between dietary habits in these areas, with an increased frequency of consuming uncooked vegetation an important risk factor for infection. Other infected animals may act as a reservoir for the parasites. Human infection is considered a serious health problem in areas such as Bolivia and neighbouring Andean countries, Egypt, especially in the area of the Nile delta, Cuba, and Iran [4].

Current treatment for *F. hepatica* infection often relies on the drug triclabendazole, which is believed to disrupt microtubule formation since this is how the related family of compounds, the benzimidazoles, act against nematodes [5]. Evidence for this mode of action of triclabenazole has been provided by the inhibition of microtubule dependent processes when flukes are incubated with the drug. These processes include mitosis of sperm cells and viteline cells [6,7], and the migration of secretion bodies through the tegument [8]. However, isolates that show resistance to triclabendazole have been identified, first in Australia and now throughout Europe including the British Isles [9]. This resistance is complicated by an increase in the incidence of infection, possibly caused by wetter, milder winters expanding the habitat of the intermediate snail host. As no new flukicidal drugs have been developed in more than 20 years, it is important that new drugs and new drug targets are identified and developed. Vaccination is another option to prevent infection, and although there is no current vaccine available, there is evidence that certain antigens derived from *F. hepatica* may confer protection against infection by the liver fluke. Vaccines would ideally target new excysted juveniles and immature flukes, since the migrations of the juvenile flukes is important in the pathology of infection. There is evidence that infected cattle naturally acquire resistance against further infection and understanding how this natural immunity would help in the development of an effective vaccine. This would involve determining which antigen or antigens are involved and which branch of the hosts' immune system is involved [10].

FASCIOLA HEPATICA CALCIUM BINDING PROTEINS

The EF-hand is a short sequence motif which is acts as the calcium ion binding site in a number of diverse proteins [11]. A number of EF-hand-containing proteins have been characterised from *F. hepatica* and these fall into three key families – the 22 kDa EF-hand/dynein light chain proteins, the calmodulin-like proteins and 8 kDa EF-hand proteins. Proteins from these three families have also been described from other, related trematodes. In many cases the roles of these proteins in the organism are unclear; however, it seems likely that they play roles in the transduction of calcium signalling. Disruption of calcium signalling is likely to be deleterious to the liver fluke and so the selective antagonism of these proteins may offer a route to novel therapeutic agents.

22 kDa EF-Hand/Dynein Light Chain Proteins

The first *F. hepatica* protein to be identified in this family was FH22. This is a 190 residue protein which contained two putative EF-hands and was estimated to have a molecular weight of 22 kDa [12]. The sequence similarity to human calmodulin was only 16 %. When analysing the sequences of the

EF-hand motifs as compared to the consensus sequence it was found that only the second EF-hand was in complete agreement. The first EF-hand was in agreement with the residues that directly involved with coordinating with the calcium ion but it lacked the important glycine at position 6 in the loop, which allows the loop to turn sharply enough to allow the other chelating residues to be in the correct position to do so. Recombinant FH22 was expressed and purified from *Escherichia coli* and its ability to bind calcium ions was assessed by determining changes in electrophoretic mobility in native (i.e. containing no SDS or other denaturing agents) polyacrylamide gel in the presence and absence of calcium ions. A retardation in the sample containing calcium compared to the sample that contained no calcium ions indicated that FH22 could bind calcium ions [12]. This retardation is believed to happen because the bound calcium ion, which is positively charged, is attracted to the anode as well as masking the negative charges of the EF-hand so that it slows the progression of the negatively charged FH22 towards the cathode. The conformational change caused by the binding of calcium is also likely to affect on the electrophoretic mobility.

A sequence from a *F. gigantica* cDNA library was found which coded for a protein with 96% identity to FH22. This protein, FgCaBP1 contains two predicted EF-hand motifs and changes in electrophoretic mobility indicated that it also bound calcium ions [13]. When FgCaBP1 was localised using both *in situ* RNA hybridisation and immunolocalisation, it was found in the tegument of both adult and juvenile flukes. The antibody raised against recombinant FgCaBP1 reacts with a 26 kDa antigen from sera from mice infected with *Schistosoma japonicum* and mice infected with *Schistosoma mekongi* [13]. This would suggest that there is a family of these tegumental proteins in the digenean trematodes.

Recent investigations have demonstrated that there at least two more proteins from this family in *F. hepatica*. Both of these proteins, FhCaBP3 and FhCaBP4, have highly similar homologues in *F. gigantica* − FgCaBP3 and FgCaBP4 respectively [14-16]. FhCaBP3 and FhCaBP4 both bind calcium and manganese (but not magnesium) ions as judged by native polyacrylamide gel electrophoresis. They are also both able to homodimerise. However, they differ in how calcium ions regulate this dimerisation event. In the case of FhCaBP3 dimerisation is favoured in the absence of calcium ions, whereas dimerisation of FhCaBP4 is stimulated by calcium ions [15,16]. The modelled structures of FH22, FhCaBP3 and FhCaBP4 reveal common features. All are divided into two domains separated by a flexible linker. The N-terminal domain is largely α-helical and contains the two EF-hands. The C-terminal, dynein light chain-like domain is predominantly β-sheet [15,16] (Figure 1). As yet, the molecular mechanism(s) of dimerisation is unknown. In FhCaBP3, binding of calcium (or manganese) ions to the second EF-hand is predicted to result in loss of β-sheet structure in the C-terminal domain. This suggests that ion binding results in conformational changes which are transmitted from one domain to the other [16].

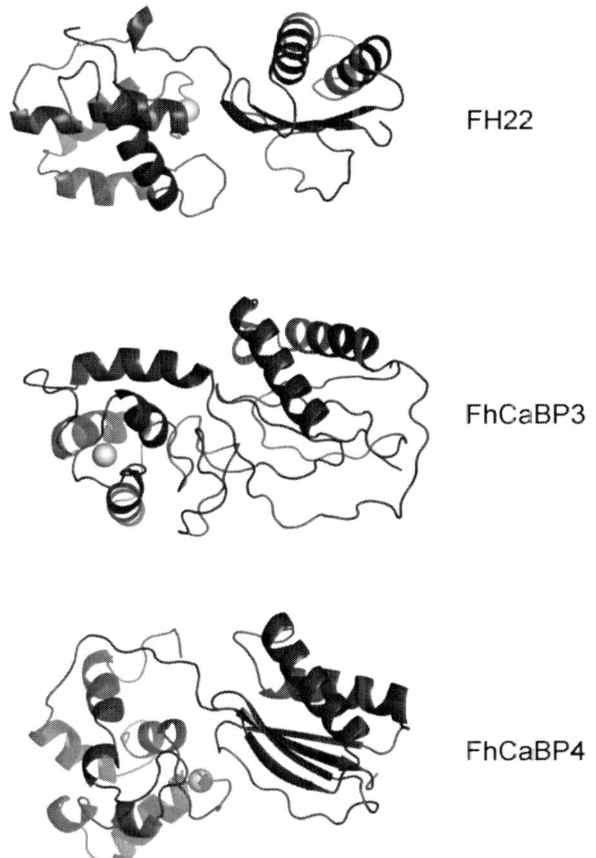

Figure 1. The predicted structures of FH22, FhCaBP3 and FhCaBP4. The structure of FH22 was modelled using Phyre2 [17], a calcium ion inserted into the second EF-hand by alignment with the Reps1 EH domain (PDB 1FI6 [18]) and the structure then computationally solvated and energy minimised using Yasara [19]. Models of FhCaBP3 and FhCaBP4 have been previously reported [15,16]. The images were produced using PyMol (www.pymol.org) and calcium ions are represented by spheres.

Calmodulin-Like Proteins

Russell et al. [20] identified two calmodulin-like proteins (FhCaM1 and FhCaM2) from an adult *Fasciola hepatica* EST databank each of which contained four EF-hands. FhCaM1 shows 98 % identity to mammalian calmodulin and FhCaM2 shows 40 % identity to mammalian calmodulins. By using native gel electrophoresis in the presence and absence of calcium ions, it was determined that both FhCaM1 and FhCaM2 could bind Ca^{2+} ions and that both proteins could be purified in a calcium-dependent manner when using phenyl-Sepharose, which demonstrated that they undergo a conformational change which exposed hydrophobic patches when bound to calcium ions [20]. This indicates that these proteins are capable of acting as calcium sensors in at least the adult flukes. Homology modelling was carried out which showed that they are predicted to adopt the standard calmodulin fold i.e. they had two globular domains each containing two EF-hands joined by a helical linker peptide [20] (Figure 2).

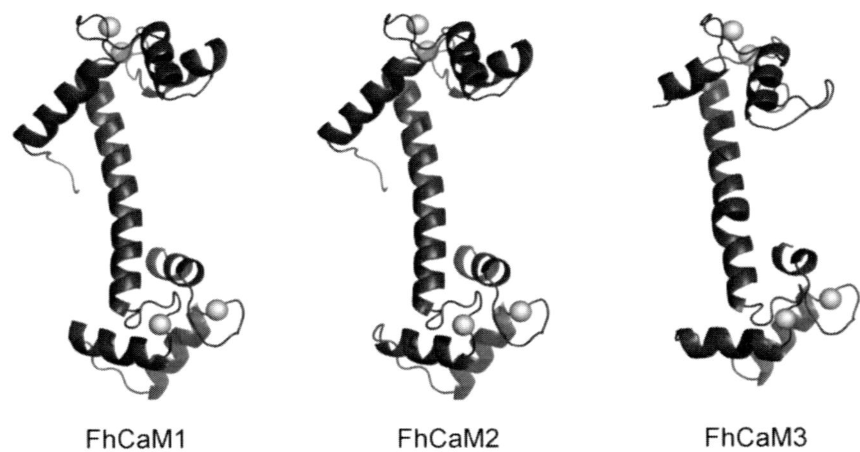

Figure 2. The predicted structures of FhCaM1, FhCaM2 and FhCaM3 in extended, calcium bound forms. Both FhCaM1 and FhCaM2 were initially modelled in Phyre2 [17] and then calcium ions were added in from the top-ranked extended calcium-bound template (in both cases, the potato calmodulin PCM6, PDB 1RFJ [21]). The calcium bound structures were the computationally solvated and energy minimised using Yasara [19]. The FhCaM3 model was previously reported [22]. The images were produced using PyMol and calcium ions are represented by spheres.

When the individual EF-hands were analysed in FhCaM1 it was determined that, based on their sequence and modelled structure, all four should be able to bind calcium ions. In FhCaM2 however, there were a number of residue substitutions that were predicted to lower affinity for calcium ions. The first of these is seen in the second EF-hand of the N-terminal globular region at position 60 in which the conserved asparagine residue is replaced with a serine. The amide oxygen of the asparagine is usually involved in the coordination of the calcium ion and this substitution to serine would lower affinity of the calcium ion as its side chain oxygen would be too distant and not polarised enough to allow the EF-hand to have the same affinity for the calcium ion. Another key change is in the second EF-hand of the C-terminal globular domain in which, at position 135, the asparagine of the canonical EF-hand sequence has been replaced with a lysine. This substitution adds a positive charge which could decrease the affinity for calcium ions to an extent that it would most likely prevent this EF-hand from coordinating a calcium ion at physiological concentrations [20].

The actual functions of FhCaM1 and FhCaM2 have not been determined but since FhCaM1 is so similar to other well-studied calmodulin proteins, it is likely that it has the same function, and would act as a calcium sensor with a broad range of protein targets. To date, the only biochemically characterised interaction is that with the plasma membrane calcium-ATPase (PMCA) [23]. FhCaM2, being quite unlike other calmodulins but still relatively similar in sequence could possibly have a specialised function within adult flukes.

Two calmodulin transcripts, SmCaM1 and SmCaM2, with 99 % identity to each other were found in an EST databank of *S. mansoni* transcripts [24]. SmCaM1 and SmCaM2 were also found to show high sequence identity with FhCaM1 (99% and 98% respectively), human calmodulin (98 % and 97 %) and *Drosophila* calmodulin (98 % and 99 %). Native gel electrophoresis showed that recombinant SmCaM1 was retarded in the presence of calcium ions. The number of miracidium to sporocyst transformations was measured after an 18 hour dose of the calmodulin antagonist trifluoperazine (TFP). It was found that doses of 5 µM and 10 µM caused a significant decrease in the number of these transformations whereas doses of 0.5 µM and 2 µM did not. RNA interference (RNAi) experiments using double stranded transcripts of SmCaM1 and SmCaM2 showed that a reduction in transcription of these sequences caused a decrease in the average sporocyst length [24].

FhCaM3 is a newly discovered calmodulin-like protein found in *F. hepatica*. In a BLAST search the most similar were the 16 kDa egg antigen protein from *S. mansoni* (SmE16) and a similar protein from *S. japonicum*.

Like the other calmodulin-like proteins, it showed retardation in native polyacrylamide gels in the presence of calcium and can be selected from a hydrophobic surface in a calcium dependent manner [22]. SmE16 has been shown to bind calcium and antibodies from the sera from animals infected with *S. mansoni* react to recombinant SmE16 demonstrating that it is secreted [25].

In molecular models of FhCaM3, the protein adopts the overall calmodulin fold with two globular domains, which contain two EF-hands each, and that these domains are connected by a helical linker (Figure 2). Although the prediction of the secondary structure of the central linker is of α-helix, there appears to be a distortion in the helix resulting in a lump so that one residue protrudes from the helix [22]. In calmodulins the central linker is flexible which allows the globular domains to wrap around the calmodulin-binding domain of the target protein. Structural studies of calmodulins show that when in the calcium bound form this central linker is helical in character; however, *in vivo* calmodulin is likely to be in association with its target protein, i.e. the target-bound form, and so the linker is likely to be kinked. The effects of this "lump" on target recognition are currently unknown.

The four EF-hands in the model of FhCaM3 all show slightly unusual sequences in which positions 1, 3, 5 and 7 and the conserved glutamate in the exiting helix are all in place to coordinate with the calcium ion as in the canonical EF-hand, but at position 9 in the loop appears to be too far away from the coordination sphere to be involved in the coordination. The residues at these positions are serines in the first EF-hand of the N-terminal globular domain and in both EF-hands of the C-terminal globular domain, and a threonine in the second EF-hand of the N-terminal globular domain which can be found in canonical EF-hands although aspartates and asparagines are more common. With these residues too far from the calcium ion to be involved its coordination, the overall affinity for the ion would be reduced which would mean that a higher concentration of calcium ions may be required to saturate the binding sites [22]. The residue at position 7 in the canonical EF-hand loop coordinates the calcium ion with its backbone oxygen, and because of this there are no constraints on which residue occupies this position. In the first EF-hand of the N-terminal domain and the second EF-hand of the C-terminal domain in FhCaM3 this residue position 7 is non-polar as a phenylalanine and an alanine respectively. These residues are unlikely to affect the affinity for the calcium ions as they have no charge. In both the second EF-hand of the N-terminal domain and the first EF-hand of the C-terminal domain, position 7 is occupied by a lysine. These lysines could reduce the affinity for the calcium ion at these positions because it will be adding a positive charge to the area around the coordination sphere.

On a native gel, FhCaM3 runs further than FhCaM2 but travels less far than FhCaM1. This demonstrated experimentally that it has a higher pI than FhCaM1 but a lower pI than FhCaM2 [22]. The Connolly surface of FhCaM3 shows that despite being generally a polar surface the areas between the globular domains and the central linker appear to be more non-polar in nature. These areas are the parts of the calmodulins which would normally interact with the target proteins and which involved hydrophobic packing. It is also the area which calmodulin-antagonists are likely to bind to. In the N-terminal domain, this area is more like those of mammalian calmodulins, in that it is more open when compared to the same area in the C-terminal domain. The area under the C-terminal domain contains a deep but narrow hydrophobic cleft which might not be wide enough to accommodate the typical target peptide.

Therefore, it could be that FhCaM3 only interacts with its target(s) via the N-terminal domain. It is also possible that when the target peptide is close to this cleft it induces a conformational change which would allow the interaction to take place [22].

FhCaM3 was localised to the vitelline cells and to the inside of eggs in the ovaries [22]. The shell of *F. hepatica* eggs are comprised of a quinone-tanned protein.

The eggshell precursor protein is synthesised in the vitelline cells of the mature fluke and has been found to contain high levels of the relatively unusual amino acid L-3,4-dihydroxyphenylalanine (L-DOPA) [26]. During the tanning process, these L-DOPA residues are converted to *o*-quinones by the action of a phenol oxidase enzyme and these then crosslink the precursors to form a tough, waterproof surface. Inside the vitelline cells, the precursor proteins form globules but the low pH prevents cross-linking. The release of the precursor protein has been shown to be calcium-dependent as treatment of the calcium ionophore lasalocid causes the proteins to be released prematurely. When the proteins are released, they form a uniform layer around the ovum and the tanning occurs [27].

Since FhCaM3 is a calcium binding protein localised to the vitelline cells and eggs of the mature fluke and that release of the eggshell precursor proteins is a calcium-dependent process, it is possible that FhCaM3 is involved in the calcium signalling in these tissues. SmE16 is the most similar protein to FhCaM3 in a BLAST search, which would suggest that they could be involved in similar processes.

The eggshell formation process of *Schistosoma* species has been found to be similar to that of *F. hepatica* including the calcium dependent release of the eggshell precursor proteins from the vitelline cells [28].

8 kDa EF-Hand Proteins

FH8 is an 8 kDa protein from *F. hepatica* which contains two EF-hand motifs. The coding sequence was found in a screen of a cDNA library and it was found to be secreted from the surface of the parasite between 1–3 weeks post infection [29]. Native gel electrophoresis showed that it bound calcium ions. Sequence alignment studies were carried out, lining up the sequence with the N- and C-terminal sequences of FhCaM1 and FhCaM2 individually, finding identities of 19 % and 26% for FhCaM1 and 22% and 33% for FhCaM2. FH8 shows more identity to a range of proteins including SmCaBP (40%), SjCa8 (37%) and Ch8 (55%) which are found in *S. mansoni, S. japonicum* and *Clonorchis sinesis* respectively [30]. Studies on SjCa8 showed that it is also located on the surface during the initial infection [31].

SjCa8 was identified from a *S. japonicum* cercarial cDNA library. It shows 83% identity with SmCaBP which had been found in cDNA from *S. mansoni* cercaria [32].

As with FH8, these proteins contain two EF-hand motifs and they show a calcium-dependent retardation during native electrophoresis. Stage-specific reverse-transcriptase PCR show that SjCa8 is only expressed in the first six hours of the cercarial stage and immunolocalisation show that the protein is found at the surface of the cercaria [31].

As with FH8 and SmCaBP, SjCa8 is also secreted from the parasite, and because of the stage at which SjCa8 can be found, it is possible that this family of proteins is involved in penetration of the host tissues, possibly in the regulation of the enzymes required.

Homology modelling of FH8 in the apo and calcium bound states gave rise to models in which there was an increase in the hydrophobic area available to the solvent when calcium bound compared to the apo structure. Measurements of the change in the fluorescence of 8-anilonaphthalene-1-sulphonate (ANS) in the presence and absence of calcium ions were carried out.

ANS interacts with hydrophobic areas of proteins and when in such an interaction its fluorescence emission intensity is increased. It was found that in the presence of calcium ions, the fluorescence emission intensity was increased.

This would suggest that the homology model was correct in predicting a conformational change that increased the hydrophobic area that is solvent accessible [30].

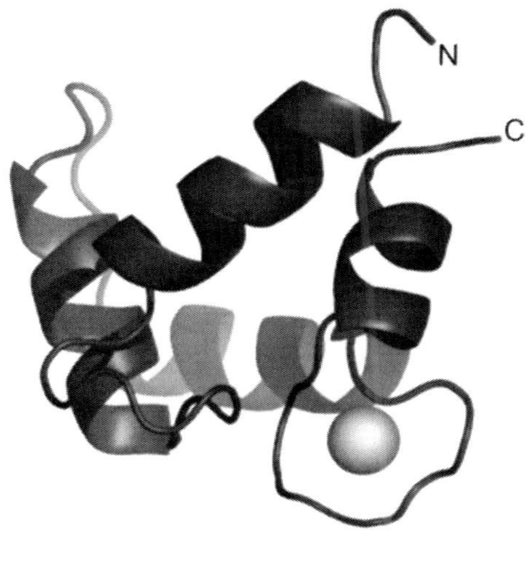

Figure 3. The predicted structure of FH8. The initial model was derived using Phrye2 [17] and the position of the calcium ion estimated by superimposing the model onto the most similar experimentally determined structure (*Chlamys nipponensis akazara* troponin C, PDB 3TZ1[33]). The model of calcium bound FH8 was solvated and energy minimised using Yasara [19]. The image was generated using PyMol and calcium ions are represented by spheres.

Conclusion

The common liver fluke contains a number of families of calcium-binding proteins. These include molecules such as FhCaM1 which are highly similar to the equivalent proteins found in the host. However, even within the calmodulin family, FhCaM2 and FhCaM3 show significant structural and biochemical differences from the mammalian host proteins. The 22 kDa EF-hand/dynein light chain and 8 kDa EF-hand families appear to be unique to trematodes. It is reasonable to speculate that these molecules are, therefore, involved in processes which are unique to this class of organisms.

Furthermore, there are biochemical variations within these families; for example the different response of FhCaBP3 and FhCaBP4 to calcium ions. This suggests that while these family members are likely to be involved in calcium sensing and signalling, the responses they mediate are likely to be different. However, at present, virtually nothing is known about the *in vivo* roles of these proteins and discovering these roles should be seen as a research priority in this field.

References

[1]   Torgerson,P. and Claxton,J. (1999) Epidemiology and control. In Dalton,J.P. (ed.), *Fasciolosis*. CAB International, Wallingford, UK, pp. 113-149.

[2]   Genicot, B., Mouligneau, F. & Lekeux, P. (1991). Economic and production consequences of liver fluke disease in double-muscled fattening cattle. *Zentralbl Veterinarmed B. 38,* 203-208.

[3]   Robinson, M. W. & Dalton, J. P. (2009). Zoonotic helminth infections with particular emphasis on fasciolosis and other trematodiases. *Phil Trans. R. Soc. Lond B. Biol. Sci. 364,* 2763-2776.

[4]   Mas-Coma, S. (2005). Epidemiology of fascioliasis in human endemic areas. *J. Helminthol. 79,* 207-216.

[5]   Lacey, E. (1988). The role of the cytoskeletal protein, tubulin, in the mode of action and mechanism of drug resistance to benzimidazoles. *Int. J. Parasitol. 18,* 885-936.

[6]   Stitt, A. W., Fairweather, I., Trudgett, A. G. & Johnston, C. F. (1992). *Fasciola hepatica*: localization and partial characterization of tubulin. *Parasitol Res. 78,* 103-107.

[7]   Stitt, A. W. & Fairweather, I. (1996). *Fasciola hepatica*: disruption of the vitelline cells in vitro by the sulphoxide metabolite of triclabendazole. *Parasitol. Res. 82,* 333-339.

[8]   Stitt, A. W. & Fairweather, I. (1994). The effect of the sulphoxide metabolite of triclabendazole ('Fasinex') on the tegument of mature and immature stages of the liver fluke, *Fasciola hepatica. Parasitology. 108 (Pt 5),* 555-567.

[9]   Chemale, G., Perally, S., LaCourse, E. J., Prescott, M. C., Jones, L. M., Ward, D., Meaney, M., Hoey, E., Brennan, G. P., Fairweather, I., Trudgett, A. & Brophy, P. M. (2010). Comparative proteomic analysis

of triclabendazole response in the liver fluke *Fasciola hepatica*. *J. Proteome Res. 9,* 4940-4951.

[10] Ross, J. G. (1967). Studies of immunity to *Fasciola hepatica*: acquired immunity in cattle, sheep and rabbits following natural infection and vaccine procedures. *J. Helminthol. 41,* 393-399.

[11] Gifford, J. L., Walsh, M. P. & Vogel, H. J. (2007). Structures and metal-ion-binding properties of the Ca^{2+}-binding helix-loop-helix EF-hand motifs. *Biochem. J. 405,* 199-221.

[12] Ruiz de Eguino, A. D., Machin, A., Casais, R., Castro, A. M., Boga, J. A., Martin-Alonso, J. M. & Parra, F. (1999). Cloning and expression in *Escherichia coli* of a *Fasciola hepatica* gene encoding a calcium-binding protein. *Mol. Biochem. Parasitol. 101,* 13-21.

[13] Vichasri-Grams, S., Subpipattana, P., Sobhon, P., Viyanant, V. & Grams, R. (2006). An analysis of the calcium-binding protein 1 of *Fasciola gigantica* with a comparison to its homologs in the phylum Platyhelminthes. *Mol. Biochem. Parasitol. 146,* 10-23.

[14] Subpipattana, P., Grams, R. & Vichasri-Grams, S. (2012). Analysis of a calcium-binding EF-hand protein family in *Fasciola gigantica*. *Exp. Parasitol. 130,* 364-373.

[15] Orr, R., Kinkead, R., Newman, R., Anderson, L., Hoey, E. M., Trudgett, A. & Timson, D. J. (2012). FhCaBP4: a *Fasciola hepatica* calcium-binding protein with EF-hand and dynein light chain domains. *Parasitol. Res. 111,* 1707-1713.

[16] Banford, S., Drysdale, O., Hoey, E. M., Trudgett, A. & Timson, D. J. (2013). FhCaBP3: A *Fasciola hepatica* calcium binding protein with EF-hand and dynein light chain domains. *Biochimie. 95,* 751-758.

[17] Kelley, L. A. & Sternberg, M. J. (2009). Protein structure prediction on the Web: a case study using the Phyre server. *Nat. Protoc. 4,* 363-371.

[18] Kim, S., Cullis, D. N., Feig, L. A. & Baleja, J. D. (2001). Solution structure of the Reps1 EH domain and characterization of its binding to NPF target sequences. *Biochemistry. 40,* 6776-6785.

[19] Krieger, E., Joo, K., Lee, J., Lee, J., Raman, S., Thompson, J., Tyka, M., Baker, D. & Karplus, K. (2009). Improving physical realism, stereochemistry, and side-chain accuracy in homology modeling: Four approaches that performed well in CASP8. *Proteins. 77 Suppl 9,* 114-122.

[20] Russell, S. L., McFerran, N. V., Hoey, E. M., Trudgett, A. & Timson, D. J. (2007). Characterisation of two calmodulin-like proteins from the liver fluke, *Fasciola hepatica*. *Biol. Chem. 388,* 593-599.

[21] Yun, C. H., Bai, J., Sun, D. Y., Cui, D. F., Chang, W. R. & Liang, D. C. (2004). Structure of potato calmodulin PCM6: the first report of the three-dimensional structure of a plant calmodulin. *Acta Crystallogr D Biol. Crystallogr. 60,* 1214-1219.

[22] Russell, S. L., McFerran, N. V., Moore, C. M., Tsang, Y., Glass, P., Hoey, E. M., Trudgett, A. & Timson, D. J. (2012). A novel calmodulin-like protein from the liver fluke, *Fasciola hepatica. Biochimie. 94,* 2398-2406.

[23] Moore, C. M., Hoey, E. M., Trudgett, A. & Timson, D. J. (2012). A plasma membrane Ca^{2+}-ATPase (PMCA) from the liver fluke, *Fasciola hepatica. Int. J. Parasitol. 42,* 851-858.

[24] Taft, A. S. & Yoshino, T. P. (2011). Cloning and functional characterization of two calmodulin genes during larval development in the parasitic flatworm *Schistosoma mansoni. J. Parasitol. 97,* 72-81.

[25] Moser, D., Doenhoff, M. J. & Klinkert, M. Q. (1992). A stage-specific calcium-binding protein expressed in eggs of *Schistosoma mansoni. Mol. Biochem. Parasitol. 51,* 229-238.

[26] Waite, J. H. & Rice-Ficht, A. C. (1987). Presclerotized eggshell protein from the liver fluke *Fasciola hepatica. Biochemistry. 26,* 7819-7825.

[27] Colhoun, L. M., Fairweather, I. & Brennan, G. P. (1998). Observations on the mechanism of eggshell formation in the liver fluke, *Fasciola hepatica. Parasitology. 116 (Pt 6),* 555-567.

[28] Wells, K. E. & Cordingley, J. S. (1991). *Schistosoma mansoni*: eggshell formation is regulated by pH and calcium. *Exp. Parasitol. 73,* 295-310.

[29] Silva, E., Castro, A., Lopes, A., Rodrigues, A., Dias, C., Conceicao, A., Alonso, J., Correia da Costa, J. M., Bastos, M., Parra, F., Moradas-Ferreira, P. & Silva, M. (2004). A recombinant antigen recognized by *Fasciola hepatica*-infected hosts. *J. Parasitol. 90,* 746-751.

[30] Fraga, H., Faria, T. Q., Pinto, F., Almeida, A., Brito, R. M. & Damas, A. M. (2010). FH8--a small EF-hand protein from *Fasciola hepatica. FEBS J. 277,* 5072-5085.

[31] Lv, Z. Y., Yang, L. L., Hu, S. M., Sun, X., He, H. J., He, S. J., Li, Z. Y., Zhou, Y. P., Fung, M. C., Yu, X. B., Zheng, H. Q., Cao, A. L. & Wu, Z. D. (2009). Expression profile, localization of an 8-kDa calcium-binding protein from *Schistosoma japonicum* (SjCa8), and vaccine potential of recombinant SjCa8 (rSjCa8) against infections in mice. *Parasitol. Res. 104,* 733-743.

[32] Hu, S., Law, P., Lv, Z., Wu, Z. & Fung, M. C. (2008). Molecular characterization of a calcium-binding protein SjCa8 from *Schistosoma japonicum*. *Parasitol. Res. 103,* 1047-1053.

[33] Kato, Y. S., Yumoto, F., Tanaka, H., Miyakawa, T., Miyauchi, Y., Takeshita, D., Sawano, Y., Ojima, T., Ohtsuki, I. & Tanokura, M. (2013). Structure of the Ca^{2+}-saturated C-terminal domain of scallop troponin C in complex with a troponin I fragment. *Biol. Chem. 394,* 55-68.

In: New Developments in Calcium … ISBN: 978-1-62948-601-7
Editor: Masayoshi Yamaguchi © 2014 Nova Science Publishers, Inc.

Chapter VI

Role of Regucalcin in Calcium Signaling of Kidney Cells

Masayoshi Yamaguchi[*]
Department of Hematology and Medical Oncology,
Emory University School of Medicne, Atlanta, GA, US

Abstract

Regucalcin was discovered in 1978 as a calcium-binding protein that does not contain EF-hand motif of calcium-binding domain, which differs from calmodulin and other calcium-related proteins. Regucalcin plays a pivotal role as a regulatory protein of calcium signaling in various cell types. Regucalcin is greatly pronounced in hepatocytes and kidney proximal tubular cells. This chapter provides recent information concerning the role of regucalcin in the regulation of kidney cell function. Regucalcin has been demonstrated to play a physiological role in the regulation of Ca^{2+} homeostasis in the kidney proximal tubular epithelial cells, which participate in transcellular transport to reabsorption of calcium from filtrated urinary calcium. Regucalcin has also been shown to regulate various enzyme activities including Ca^{2+}-dependent protein kinases and protein phosphatases, nitric oxide synthase, cyclic adenosine

---

[*] Corresponding author: Masayoshi Yamaguchi, Department of Hematology and Medical Oncology, Emory University School of Medicne, Atlanta, GA, US. E-mail: yamamasa1155 @yahoo.co.jp.

monophosphate phosphodiesterase, and others, which are involved in intracellular signalling pathways.

Regucalcin is localised to the nucleus and regulates the gene expressions of proteins, which are related to mineral transport-related proteins, Smad 2, the nuclear factor-kappa B, cell proliferation and apoptosis-related proteins in the kidney tubular epithelial cells. Regucalcin reveals a suppressive effect on cell proliferation and apoptotic cell death that are induced through various signalling stimulators. Moreover, regucalcin gene expression is suppressed with hypertensive state or the development of drug-induced kidney disorder. This chapter introduces recent findings that regucalcin plays a pivotal role as a multifunctional protein in the regulation of kidney proximal tubular epithelial cells.

Introduction

Calcium signals mediate many intracellular responses, which are amplified through calmodulin and protein kinase C [1-3]. Regucalcin was discovered in 1978 as a calcium-binding protein that does not contain EF-hand motif of calcium-binding domain, which differs from calmodulin and other calcium-related proteins [4].

The name, regucalcin, is proposed for this calcium-binding protein, which suppresses various Ca^{2+}- or Ca^{2+}/calmodulin-dependent enzyme activations [5-7]. Regucalcin and its gene (*rgn*) are identified in over 15 species consisting of regucalcin family and is highly conserved in vertebrate species throughout evolution [8-10].

The regucalcin gene is localized on X chromosome [11, 12]. The organisation of the regucalcin gene consists of seven exons and six introns [13]. Various transcription factors (including AP-1, NF1-A1, RGPR-p117, β-catenin, SP1 and others) are identified as the enhancer and suppressor for regucalcin gene expression [8, 14-19]. mRNA expression of regucalcin and its protein content are pronounced in the liver and kidney cortex of rats [14, 20], and regucalcin gene expression is regulated through various hormonal stimulations and physiological states [8].

The role of regucalcin in cellular function is focused in the liver in detail [21, 23]. Regucalcin plays a role in the maintenance of intracellular calcium ion (Ca^{2+}) homeostasis [24], depression of calcium signalling from the cytoplasm to the nucleus in proliferating cells [25] and regulation of nuclear functions [26, 27]. Overexpression of endogenous regucalcin in liver cells suppresses cell proliferation [28] and apoptotic cell death [29], which is induced by various signalling factors.

Regucalcin has been proposed to play a pivotal role as a suppressor protein in the regulation of cellular function in maintaining cell homeostasis that is mediated through various signalling pathways. In addition, the protein named as senescence marker protein-30, which is identical to regucalcin, was also reported after the discovery of regucalcin [30, 31].

Regucalcin is predominantly expressed in the liver and kidney cortex including kidney proximal tubular epithelial cells [14, 20]. The role of regucalcin is elucidated in liver cells [21, 22]. In recent years, there has been growing evidence that regucalcin has an important role in the regulation of kidney cell function and that regucalcin plays a pathophysiological role in renal failure. This review has been written to outline the recent advances that have been made concerning the regulation of regucalcin gene expression and the role of regucalcin in the kidney proximal tubular epithelial cells.

Horonal Regulation of Regucalcin Gene Expression

Regucalcin is expressed in the kidney tissues of rats [14, 20]. The concentration of regucalcin in rat kidney tissues has been shown to be in the range of 1.74–3.50×10^{-6} M in male or female rats as measured by using enzyme-linked immunoadsorbent assay [20]. This level is not lowered with aging [20].

Regulation of Regucalcin Gene Expression *In Vivo*

The expression of regucalcin mRNA is predominant in the kidney cortex but not in the medulla of rats [32]. Kidney cortex constitutes nephrons including glomerulo and tubulo. Regucalcin mRNA expression is enhanced in the kidney cortex after a single intraperitoneal administration of calcium chloride in rats. The administration of calcium (100 and 150 mg/kg body weight) causes a remarkable increase in serum calcium concentration and a significant elevation of calcium content in the kidney cortex at 20 minutes after calcium administration [32]. The expression of regucalcin mRNA in the kidney cortex is increased at 60 minutes after calcium administration [32].

These findings suggest that regucalcin mRNA expression is related to calcium that is increased in the kidney cortex after calcium administration.

Regucalcin mRNA expression is also enhanced after calcium administration in thyroparathyroidectomised rats with deficiency of calcium-regulation hormones (calcitonin and parathyroid hormone), which are secreted from thyroid-parathyroid glands [32]. Calcium-regulating hormones regulate calcium reabsorption in the kidney proximal tubular cells [33, 34]. The stimulating effect of calcium administration on kidney regucalcin mRNA expression may be independent on calcium-regulating hormones.

Calcium administration-stimulated regucalcin mRNA expression in the kidney cortex of rats *in vivo* is completely blocked after the treatment of trifluoperazine (TFP), an inhibitor of Ca^{2+}/calmodulin, suggesting that the expression is mediated through Ca^{2+}/calmodulin that is involved in the activation of protein kinases [32].

The nuclear proteins in the kidney cortex have been shown to specifically bind to the 5'-flanking region of rat regucalcin gene and its binding activity is partly mediated through the Ca^{2+}/calmodulin-signalling pathway [35]. This specific nuclear factor binds to the NF1 consensus motif $TTGGC(N)_6 CC$ in the promoter region of regucalcin gene in the rat kidney cortex [36]. NF1 binding motif in the promoter region of regucalcin gene may be partly involved in the transcriptional regulation of regucalcin gene expression in the kidney cortex cells.

Steroid hormones have also been shown to regulate regucalcin mRNA expression in the kidney cortex of rats. Regucalcin mRNA expression is stimulated after a single subcutaneous administration of dexamethasone (1 mg/ kg body weight) in rats, while it is suppressed after a single subcutaneous administration of aldosterone (25–100 µg/kg) or oestrogen (0.5–2 mg/kg) [37]. The administration of hydrocortisone (5–30 mg/kg) does not have an effect on RGN mRNA expression in the kidney cortex [37]. The regucalcin gene may have response elements for receptors of glucocorticoid, aldosterone or oestrogen.

The dexamethasone-induced increase in regucalcin mRNA expression in the kidney cortex is completely suppressed after simultaneous administration of cycloheximide, an inhibitor of protein synthesis [37]. This finding may be partly based on newly synthesised DNA binding proteins (steroid receptors) in the kidney cortex.

Regucalcin mRNA expression in the kidney cortex of rats has been shown to attenuate in a pathophysiological state. Regucalcin mRNA expression suppresses in the kidney cortex of adrenalectomised (ADX) rats [38]. ADX may deplete endogenous steroid hormones, which are secreted from adrenal glands.

However, the treatment of dexamethasone in ADX rats does not restore ADX-induced suppression of RGN mRNA expression in the kidney cortex [36]. Adrenal glands, however, may participate in the regulation of regucalcin mRNA expression in the kidney cortex of rats.

Regucalcin mRNA expression has been shown to be suppressed in the kidney cortex of normal or ADX rats upon oral intake of saline for 7 days, which may cause a hypertensive condition [38-40]. The increase in renal sodium transport following saline intake may suppress regucalcin mRNA expression in the kidney cortex of rats. Regucalcin mRNA expression is mediated through Ca^{2+}/calmodulin-dependent protein kinase, which is suppressed by saline intake in the kidney cortex [40].

The intake of saline has been shown to cause an alteration in calcium metabolism, which is related to calcium transport in the kidney cortex of rats [39]. Saline intake for 7 days in rats causes a significant increase in serum calcium concentration and an elevation of Ca^{2+}-ATPase activity in the basolateral membranes of the kidney cortex and a corresponding increase in renal cortex calcium content [39]. Saline intake-induced hypercalcemia may partly result from the tubular reabsorption of urinary calcium. Blood urea nitrogen (BUN) concentration was increased in rats with saline intake for 7 days, suggesting that saline intake causes renal disorder [39]. Interestingly, regucalcin mRNA expression is suppressed in the kidney cortex of spontaneous hypertensive rats (SHR) [38]. Saline intake, which may generate a hypertensive condition, causes a suppression of regucalcin mRNA expression in the kidney cortex of rats [36]. The suppressed regucalcin gene expression may have a role in the development of renal hypertension.

Various drugs are known to have a side effect of causing renal damage. Depending on the chemical, the mechanism of nephrotoxicity may involve direct interference with tubular or mitochondrial transport processes, covalent modification of critical cellular constituents or generation of free radicals. Cisplatin, a nephrotoxic antitumor drug [41], or cephaloridine, a nephrotoxic cephalosporin antibiotic [42], is known to change the thiol status in the renal cortex before the development of significant morphological changes. Regucalcin mRNA expression in the kidney cortex is markedly suppressed at 24 hours after a single intraperitoneal administration of cisplatin (2.5–10 mg/ kg body weight) or cephaloridine (250–1000 mg/kg body weight) in rats *in vivo* [43]. These chemical administrations induce kidney damage [43]. The suppression of regucalcin mRNA expression may have a pathophysiological role in the development of kidney damage. The binding of nuclear factor on the 5'-flanking region of the regucalcin gene promoter in the kidney cortex is suppressed after a single intraperitoneal administration of cisplatin (10 mg/kg body weight) in rats [36]. Regucalcin mRNA and its protein levels in the kidney cortex are decreased after cisplatin administration [36]. Suppression of the nuclear factor binding to the regucalcin gene promoter region may lead to a decrease in regucalcin mRNA and its protein expression in the kidney cortex of rats administered with cisplatin.

As mentioned above, regucalcin gene expression in the kidney cortex tissues in rats *in vivo* is markedly enhanced through Ca^{2+} signalling pathways and is regulated by steroid hormones. Moreover, the kidney regucalcin gene expression is suppressed with hypertensive states and with chemical toxicity that induces renal disorder, suggesting a pathophysiological role.

Regucalcin Gene Expression in Proximal Tubular Epithelial NRK52E Cells *In Vitro*

Hormonal regulation of regucalcin gene expression has been examined by using cloned normal rat kidney proximal tubular epithelial NRK52E cells *in vitro* [44]. Parathyroid hormone (PTH), 1, 25-dihydroxyvitamin D_3 or calcitonin play a role in the regulation of calcium transport in the kidney proximal tubular epithelial cells [33, 34]. Among these hormones, PTH (10 and 100 nM) has been found to stimulate regucalcin mRNA expression and its protein level in NRK52E cells [44]. PTH has a stimulatory effect on the reabsorption of calcium in the kidney proximal tubule [45]. Regucalcin plays a cell physiological role as an activator in ATP-dependent Ca^{2+} pumps in the basolateral membranes of the rat kidney cortex [46]. These findings may support the view that regucalcin plays a physiological role as a key molecule in the reabsorption of calcium in the kidney proximal tubule. The expression of regucalcin mRNA in NRK52E cells is enhanced by aldosterone (10 and 100 nM), which has a stimulatory effect on the reabsorption of sodium in the kidney proximal tubule [44]. The ATP-dependent Ca^{2+} pump and Na^+/Ca^{2+} exchange system are constituted in the basolateral membranes of the rat kidney cortex [47, 48]. It is possible that the effect of aldosterone in the kidney proximal tubule is partly mediated through regucalcin expression. In addition, regucalcin mRNA in NRK52E cells is enhanced after culture with dexamethasone (10 nmol) [44]. The effect of aldosterone or dexamethasone in the kidney proximal tubule is partly mediated through regucalcin.

Regucalcin mRNA expression has been found to be enhanced by dibutyryl cyclic adenosine monophosphate (DcAMP) (10 and 100 mM) or phorbol-12-myristate-13-acetate (PMA) (1 µM) in NRK52E cells [44]. PMA has a role as an activator of protein kinase C [2]. DcAMP may have a role as second messenger [49]. The expression of regucalcin mRNA in NRK52E cells may be partly mediated through signalling pathways that are related to cAMP or protein kinase C. The action of PTH has been known to be mediated by cAMP or inositol 1, 4, 5-triphosphate (IP$_3$)-released Ca^{2+} and protein kinase C in the cells [49]. The effect of PTH in stimulating regucalcin mRNA expression in NRK52E cells may be mediated by cAMP and/or Ca^{2+}-dependent protein kinase C in NRK52E cells. Dibucaine is an antagonist of Ca^{2+}/calmodulin-dependent protein kinase [50]. PD98059 is an inhibitor of the extracellular signal-related kinase (ERK) pathway [51]. Regucalcin mRNA expression is not altered in the presence of dibucaine or PD98059 in NRK52E cells [44], suggesting that the expression is not mediated through Ca^{2+}/calmodulin-dependent protein kinase or mitosis-activated protein (MAP) kinase, which is related to the ERK pathway. Regucalcin mRNA expression in NRK52E cells has been found to be suppressed after culture with staurosporine, an inhibitor of protein kinase C. Meanwhile, regucalcin mRNA expression is markedly enhanced after culture with PMA, an activator of protein kinase C [52]. This finding further supports the view that regucalcin mRNA expression is partly mediated through cell signalling that is related to protein kinase C in NRK52E cells. Culture with vanadate, which is an inhibitor of protein tyrosine phosphatise [53], did not cause a significant alteration in the rgucalcin mRNA expression in NRK52E cells [54], suggesting that the expression is not involved in cell signalling that is related to protein tyrosine phosphatase. Interestingly, regucalcin mRNA expression is suppressed after culture with tumor necrosis factor-α (TNF-α) or transforming growth factor-β (TGF-β) in NRK52E cells [55]. These factors, which repress regucalcin mRNA expression, have been poorly understood.

NF1-A1 and RGPR-p117 Are Involved in Enhancement of Regucalcin Gene Expression

NF1-A1 and RGPR-p117 are identified as hepatic nuclear factors that bind to the TTGGC sequence of the rat regucalcin gene promoter region using a yeast one-hybrid system [18, 56].

Overexpression of RGPR-p117 has been shown to enhance the rat regucalcin gene promoter activity that is involved in the TTGGC sequence in NRK52E cells [57, 58]. NF1-A1 enhances the regucalcin gene promoter activity, which is related to the TTGGC motif, and this enhancement is not revealed in the mutant with deletion of the $TTGGC(N)_6$ CC sequence in NRK52E cells [59]. NF1-A1 has been shown to localize in the nuclei of NRK52E cells and increase the regucalcin gene promoter activity in the cells [59].

This increase is enhanced after culture with Bay K 8644, an agonist of calcium entry into cells, or PMA, an activator of protein kinase C [59]. Ca^{2+}-dependent protein kinases may be involved in the enhancement of the regucalcin gene promoter activity in NRK52E cells.

In addition, the increase in the regucalcin gene promoter activity in NRK52E cells is completely blocked after culture with dibucaine, sturosporine, PD98059, vanadate or okadaic acid, which are inhibitors of various Ca^{2+}-dependent protein kinases and protein phosphatises [59]. The regucalcin gene promoter activity in NRK52E cells may be enhanced through various intracellular signalling factors including Ca^{2+}-dependent protein kinases, MAP kinase and protein phosphatases.

RGPR-p117 increases the rat regucalcin gene promoter activity in NRK52E cells [57, 58]. Co-transfection with NF1-A1 and RGPR-p117 does not significantly enhance the RGPR-p117-increased regucalcin promoter activity in NRK52E cells [57]. NF1-A1 or RGPR-p117 is localised in the nuclei of NRK52E cells. These transcription factors may independently regulate in the enhancement of the rat regucalcin gene promoter activity in NRK52E cells [57].

Various signalling factors are involved in the enhancement of the rat regucalcin gene promoter activity in NRK52E cells overexpressing NF1-A1 or RGPR-p117.

This enhancement seems to be mediated through protein phosphorylation and dephosphorylation in NRK52E cells. NF1-A1 or RGPR-p117, however, acts independently in the enhancement of the rat regucalcin gene promoter activity in NRK52E cells. NF1-A1 or RGPR-p117 plays a role as a transcription factor (enhancer) in the rat regucalcin gene promoter activity.

Regucalcin Regulates Intracellular Calcium Homeostasis

Kidneys plays a physiological role in the regulation of calcium homeostasis in blood by reabsorption of urinary calcium [33, 34, 47].

Renal cortex cells constituting the proximal tubular epithelial cells may play a role in the reabsorption of urinary calcium. Regucalcin mRNA is expressed in the kidney cortex but not in the medulla of rats and its expression is stimulated by calcium administration *in vivo* [32]. Regucalcin may play a role in the regulation of calcium transport in the renal proximal tubular epithelial cells.

Active Ca^{2+} reabsorption is transcellularlly transported and Ca^{2+} pumps in the basolateral membranes have to be involved to overcome the step of energy barriers at the peritubular cell side [47]. The regulation of intracellular Ca^{2+} homeostasis is important in the promotion of intracellular Ca^{2+} transport. The low cytoplasmic Ca^{2+} concentration of living cells is maintained through energy-requiring pumps. These pumps either remove Ca^{2+} to the extracellular space by transporting it across the plasma membrane or accumulate it inside intracellular organelles such as the mitochondria and endoplasmic reticulum. Intracellular Ca^{2+} homeostasis is regulated through plasma membrane (Ca^{2+}-Mg^{2+})-adenosine 5'-triphosphatase (ATPase), microsomal Ca^{2+}-ATPase, mitochondrial Ca^{2+} uptake and nuclear Ca^{2+} transport in the cells. Regucalcin has been shown to regulate Ca^{2+}-transporting systems in the renal cortex cells.

The Ca^{2+}-ATPase system exceeds the capacity of the Na^+/Ca^{2+} exchanger and plays a primary role in Ca^{2+} homeostasis of rat kidney cortex cells [47]. Regucalcin has been found to play a role as an activator of the ATP-dependent Ca^{2+} pumps (Ca^{2+}-ATPase) in the basolateral membranes isolated from the rat kidney cortex; regucalcin (100 and 1000 nM) increased Ca^{2+}-ATPase activity and stimulated [45] Ca^{2+} uptake by the basolateral membranes *in vitro* [46]. This effect of regucalcin is not seen in the presence of digitonin, which can solubilise the lipids of the plasma membranes in an enzyme reaction mixture [46]. Also, the effect of regucalcin on Ca^{2+} pump enzyme activity is completely inhibited in the presence of vanadate, an inhibitor of phosphorylation of ATPase, or *N*-ethylmaleimide, a SH-group modifying reagent [46]. Regucalcin may bind to the lipids at the close site of the Ca^{2+} pump enzyme in the basolateral membranes of the rat kidney cortex and may activate the enzyme by acting on the SH-group of the enzyme. The effect of regucalcin in increasing Ca^{2+}-ATPase activity in the basolateral membranes of the rat kidney cortex is abolished in the presence of DcAMP (0.01–1 mM), while IP_3 (0.1–10 µM) did not have an appreciable effect. DcAMP or IP_3 itself did not have an effect on Ca^{2+}-ATPase activity. Thus, the activating effect of regucalcin on the Ca^{2+}-ATPase may be regulated by intracellular cAMP.

Regucalcin has also been shown to increase Ca^{2+}-ATPase activity and ATP-dependent calcium uptake in the microsomes isolated from the rat kidney cortex *in vitro* [60]. The effect of regucalcin is completely inhibited in the presence of vanadate, an inhibitor of phosphorylation of ATPase, or *N*-ethylmaleimide, a SH-group modifying reagent. Regucalcin may increase Ca^{2+}-ATPase activity by binding to the SH-group of active sites of the enzyme and stimulating the phosphorylation of the enzyme in the microsomes of the rat kidney cortex [60]. The kidney cortex microsomal Ca^{2+}-ATPase activity and ATP-dependent Ca^{2+} uptake is inhibited by DcAMP or IP3 [60]. Calmodulin, a modulator of Ca^{2+} signalling, had a stimulating effect on microsomal Ca^{2+}-ATPase activity, although the stimulating effect of calmodulin is lower than that of regucalcin. Both proteins may be important as an activator in the microsomal ATP-dependent Ca^{2+} sequestration [60].

An ATP-dependent Ca^{2+} uptake system (Ca^{2+} uniporter) exists in the mitochondria of the kidney cortex of rats [33]. Regucalcin (50–250 nM) has been shown to stimulate Ca^{2+}-ATPase-related Ca^{2+} uniporter activity in the mitochondria [61].

The effect of regucalcin is completely blocked by ruthenium red or lanthanum chloride, which is a specific inhibitor of Ca^{2+} uniporter in the mitochondria [61], suggesting that regucalcin stimulates Ca^{2+}-ATPase-related Ca^{2+} uniporter activity in renal cortex mitochondria. Regucalcin may bind to the membranous lipids of renal cortex mitochondria and act on the SH-groups, which are active sites of Ca^{2+}-ATPase [61]. Calmodulin or DcAMP does not modulate the effect of regucalcin in increasing mitochondrial Ca^{2+}-ATPase activity [61].

The reabsorption of urinary calcium is promoted through transcellular Ca^{2+} transport in the renal proximal tubular epithelial cells [45]. Regucalcin stimulates Ca^{2+} transport (efflux) across the basolateral membranes of the renal cortex [46], the microsomal ATP-dependent Ca^{2+} sequestration [60] and the mitochondrial ATP-dependent Ca^{2+} uptake to maintain intracellular Ca^{2+} concentration [61] in the rat renal cortex.

Thus, regucalcin may play a physiological role in the regulation of intracellular Ca^{2+} homeostasis in the renal proximal tubular epithelial cells due to activating ATP-dependent Ca^{2+}-transport systems in the basolateral membranes, microsomes and mitochondria. Regucalcin may promote Ca^{2+}

reabsorption, which is based on ATP-dependent transcellular transport of Ca^{2+}, in the proximal tubular epithelial cells of the nephron tubule of the kidney cortex.

Moreover, regucalcin may play a physiological role in the regulation of calcium homeostasis in the blood through reabsorption of urinary calcium in the kidney.

Regucalcin Regulates Signaling Pathway-Related Enzyme Activity

Multifunctional Ca^{2+}/calmodulin-dependent protein kinases play an important role in the response of many cells for a calcium signal [62, 63]. Regucalcin (0.01–1 μM) has been found to have an inhibitory effect on the activation of Ca^{2+}/calmodulin-dependent protein kinase in the cytoplasm of the rat kidney cortex *in vitro* [64]. The effect of regucalcin in inhibiting Ca^{2+}/calmodulin-dependent protein kinase activity in the cytoplasm of the renal cortex is also seen at the greater concentration of added calcium chloride (100–1000 μM) in an enzyme reaction mixture [64], suggesting that the inhibitory effect of regucalcin on the enzyme activity does not mainly result from the binding of Ca^{2+} by regucalcin. Moreover, the inhibitory effect of regucalcin is seen in the presence of higher concentrations of calmodulin (4–20 μg/ml) in an enzyme reaction mixture [64]. Regucalcin may have a direct inhibitory effect on Ca^{2+}/calmodulin-dependent protein kinase in renal cortex cytosol. It is possible that regucalcin has a partial inhibitory effect on the binding of Ca^{2+}-calmodulin to the enzyme.

Protein kinase C is a diacylglycerol-activated Ca^{2+}- and phospholipids-dependent protein kinase that is widely distributed in the body [2]. Protein kinase C is capable of phosphorylating many proteins. Protein kinase C is an important enzyme, which is related to Ca^{2+} signalling in many cell types. Regucalcin has been found to inhibit protein kinase C activity in the cytoplasm of the rat kidney cortex *in vitro* [65]. The inhibitory effect of regucalcin (0.01–1 μM) on protein kinase C activity in the presence of Ca^{2+}, phosphatidylserine and dioctanoylglycerol is not seen in a reaction mixture without Ca^{2+} addition [65].

Moreover, regucalcin has an inhibitory effect on protein kinase C activity raised by addition of diacylglycerol or 12-myristate 13-acetate (PMA), which can directly activate the enzyme, in the presence of both Ca^{2+} and phosphatidylserine [66]. Regucalcin may bind to protein kinase C, and regucalcin binding-induced conformational alteration of the enzyme may inhibit the binding of Ca^{2+} to the kinase.

Thus, regucalcin may play a pivotal role as a suppressor protein in the regulation of Ca^{2+} signalling-dependent cellular functions, which are mediated through Ca^{2+}/calmodulin-dependent protein kinase and protein kinase C, in the kidney cortex cells.

Protein phosphorylation-dephosphorylation is a universal mechanism by which numerous cellular events are regulated [52]. It has become apparent that there may exist many phosphatases that, like the kinases, are elaborately and rigorously controlled [52]. Protein phosphatase plays an important role in intracellular signal transduction due to hormonal stimulation. Calcineurin, a calmodulin-binding protein, possesses a Ca^{2+}-dependent and calmodulin-stimulated protein phosphatase activity [66]. Calcineurin is a protein serine/threonine phosphatase. Regucalcin (0.01–1 µM) has been found to have an inhibitory effect on Ca^{2+}/calmodulin-dependent phosphatase (calcineurin) activity in rat renal cortex cytoplasm *in vitro* [66]. The inhibitory effect of regucalcin on Ca^{2+}/calmodulin-dependent protein phosphatase activity in the renal cortex cytoplasm is not weakened by the addition of higher concentrations of calcium chloride [67]. In addition, regucalcin does not have an effect on protein phosphatase activity in the presence of Ca^{2+}-chelator without calmodulin, suggesting that the effect of regucalcin may not result from a direct action on the enzyme [67]. Regucalcin can bind to calmodulin in an experiment using calmodulin-agarose beads *in vitro* [68]. The inhibitory effect of regucalcin on Ca^{2+}/calmodulin-dependent protein phosphatase activity in the renal cortex cytosol may be partly based on its binding to calmodulin. It cannot exclude, however, that regucalcin directly binds the enzyme.

Regucalcin (50–250 nM) has also been shown to have an inhibitory effect on Ca^{2+}/calmodulin-independent protein tyrosine phosphatase activity in rat renal cortex cytosol, although the addition of regucalcin into an enzyme reaction mixture does not have an inhibitory effect on protein phosphatase activity toward phosphoserine and phosphothreonine [69]. However, endogenous regucalcin has been shown to have an inhibitory effect on Ca^{2+}/calmodulin-increased protein phosphatase activity toward three phosphoamino acids [69]. The presence of anti- regucalcin monoclonal antibody in an enzyme reaction mixture is found to cause an increase in protein phosphatase activity toward three phosphoamino acids in the renal cortex cytosol [69], suggesting that the endogenous regucalcin, which is present in the cortex cytoplasm, has a suppressive effect on the activity of various protein phosphatases in the kidney cortex cells. Protein phosphatase activity is present in the nuclei of kidney cortex cells [70]. Ca^{2+}/calmodulin-dependent protein phosphatase (calcineurin) and protein tyrosine phosphatase may be present in the nucleus of rat kidney cortex cells [70]. Regucalcin is localized in the nucleus of the rat kidney cortex [70]. Regucalcin (25–100 nM) has an inhibitory effect on protein tyrosine phosphatase and protein serine/threonine phosphatase activities in the nucleus of renal cortex cells [70].

The endogenous regucalcin in the nucleus of kidney cortex cells suppresses various protein phosphatases, which are present in the nucleus, using regucalcin monoclonal antibody [70].

Regucalcin mRNA levels in the kidney cortex are decreased after saline intake for 7 days in rats [38]. Regucalcin in the cytoplasm and nucleus of the kidney cortex is decreased after intake of saline for 7 days in rats [69, 70]. The intake of saline causes a remarkable decrease in protein phosphatase activity in the cytoplasm and nuclei [69, 70]. In addition, the effect of anti-regucalcin monoclonal antibody in increasing protein phosphatase activity in the cytoplasm and nuclei is weakened after saline intake [69, 70], supporting the involvement of endogenous regucalcin in the regulation of protein phosphatase activity.

A single intraperitoneal administration of calcium (25–100 mg/kg body weight) in rats has been found to produce an increase in calcium content and a corresponding elevation of RGN in the cytoplasm and nucleus of the kidney cortex [71]. Calcium administration causes an increase in protein phosphatase activity in the cytoplasm and nucleus of the kidney cortex [71], suggesting that the increase in the enzyme activity is partly mediated through Ca^{2+} signalling, which results from the augmentation of renal calcium content. The presence of anti-regucalcin monoclonal antibody in the enzyme reaction mixture causes an increase in protein phosphatase in the cytoplasm and nucleus of the normal rat kidney cortex [71]. This effect is significantly enhanced in the cytoplasm and nucleus of the kidney cortex in calcium-administered rats [71]. This finding demonstrates that the calcium administration-induced increase in endogenous regucalcin suppresses the enhancement of protein phosphatase activity in the cytoplasm and nucleus of the kidney cortex in calcium-administered rats.

Dephosphorylation of many phosphorylated proteins is regulated by protein phosphatase in various cells [52]. Protein phosphatase may be implicated in transcriptional regulation in the nucleus of cells. Regucalcin can inhibit protein phosphatase activity in the nucleus of the rat kidney cortex [70, 71], suggesting that regucalcin plays an important role in the regulation of nuclear signalling, which is related to the gene expression in the nucleus of renal cortex cells.

cAMP is generated through the activation of the plasma membrane adenylate cyclase due to hormonal stimulation in many cell types and activates cAMP-dependent protein kinase, which plays an important role in cAMP signalling pathway [48]. cAMP is degraded by cAMP phosphodiestrase, which is activated by Ca^{2+}/calmodulin [1].

Regucalcin inhibits Ca^{2+}/calmodulin-dependent cAMP phosphodiesterase activity in the cytosol of the rat renal cortex [72]. Regucalcin may play a role in the regulation of both cAMP-dependent and Ca^{2+}-dependent signalling pathways that are modulated through hormonal stimulation in the renal cortex cells.

Nitric oxide (NO) acts as a messenger or modulator molecule in many biological systems [73]. NO has an unpaired electron that reacts with proteins and targets primarily through their thiol or heme groups, and it acts as a messenger or modulator molecule in many biological systems. NO is produced from L-arginine with L-citrulline as a co-product in a reaction catalysed by NO synthase that requires Ca^{2+}/calmodulin [73]. Regucalcin has been shown to inhibit NO synthase activity in the cytoplasm of the rat liver [74]. Regucalcin also has an inhibitory effect on Ca^{2+}/calmodulin-dependent NO synthase activity in the rat renal cortex cytoplasm [75], suggesting that regucalcin has a suppressive effect on overproduction of NO in the cells.

The endogenous regucalcin may depress NO synthase activity in the kidney cortex cytoplasm of regucalcin transgenic rats as compared with that of wild-type rats [75]. The effect of calcium chloride (10 μM) in increasing NO synthase activity in the kidney cortex cytosol of wild-type rats is weakened in that of regucalcin transgenic rats [75]. The presence of anti-regucalcin monoclonal antibody (25 or 50 ng/ml) in the reaction mixture caused a significant increase in NO synthase activity. This increase completely disappeared after the addition of regucalcin (100 nM) [75]. Endogenous regucalcin has a suppressive effect on NO synthase activity in the kidney cortex cytoplasm of rats.

Regucalcin has been demonstrated to have a suppressive effect on Ca^{2+}/calmodulin-dependent NO synthase activity in the kidney cortex. NO acts as a messenger or modulator in kidney cortex cells. Regucalcin may play a role as a suppressor protein in NO production in the kidney cells and may regulate many cellular events that are involved in NO signalling. NO production may be stimulated through Ca^{2+} signalling due to hormonal stimulation in kidney cortex cells. Regucalcin may have a suppressive effect on overproduction of NO in kidney cortex cells.

As mentioned above, regucalcin may play a role as a suppressor for activation of many enzymes which are related to signalling pathways that are mediated through calcium, cAMP, NO, Ca^{2+}-dependent protein kinases and protein phosphatases in the kidney cortex cells. The mechanism by which regucalcin has an inhibitory effect on Ca^{2+}/calmodulin-dependent enzyme activity may be partly based on its binding to Ca^{2+}/calmodulin and/or enzyme.

Regucalcin has been demonstrated to bind calmodulin in analysis with sodium dodecyl sulfate-polyacrylamide gel electrophoresis (SDS-PAGE) using calmodulin-agarose beads [68]. Regucalcin may directly bind to Ca^{2+} and/or enzyme to reveal the inhibitory effect on enzyme activity.

Interestingly, regucalcin has been found to stimulate proteolytic activity in the cytoplasm of the rat kidney [76, 77]. Regucalcin uniquely activates thiol proteases independent of Ca^{2+} in the cytoplasm of the rat kidney cortex, although it does not have an effect on serine proteases and metalloproteases [76, 77]. The effect of regucalcin (10–250 nM) in increasing proteolytic activity in the cytoplasm of the rat kidney cortex is seen in the presence of ethyleneglycol bis (2 aminoethylether)-N,N,N',N'-tetraacetic acid (EGTA), a chelator of Ca^{2+}, although Ca^{2+} addition-increased proteolytic activity was completely abolished by the addition of chelator [75], suggesting that regucalcin increases proteolytic activity independent of Ca^{2+} in the kidney cortex cytoplasm. Regucalcin directly acts on the SH-groups of protease in the kidney cortex cytoplasm. The molecular mechanism by which regucalcin activates thiol proteases (calpains) in the kidney cortex cytoplasm remains to be elucidated. Regucalcin activates protease at concentrations of 10–250 nM in *in vitro* experiments [76]. The concentration of regucalcin present in rat kidney tissues is about 5.3 µM [20]. Regucalcin may play a physiological role in the activation of thiol proteases in renal cortex cells.

Calpains are thiol (SH) proteases [78]. Calpains are ubiquitous, non-lysosomal, calcium-dependent proteases that may play important roles in Ca^{2+}-mediated intracellular processes [78, 79]. The ability of calpains to alter limited proteolysis, activity or function of numerous cytoskeletal proteins, protein kinases, receptors and transcription factors suggests its involvement in various Ca^{2+}-regulated cellular functions [79]. Regucalcin increases the activity of thiol proteases including calpains in the cytoplasm of the rat kidney cortex. Regucalcin may play a pivotal role in the regulation of cellular functions related to Ca^{2+} that is mediated through thiol proteases. Presumably, regucalcin plays a role in the regulation of signal transduction that is involved in proteases.

Stimulatory effect of regucalcin on proteolytic activity is impaired in the kidney cortex cytoplasm of rats with the intake of saline for 7 days [77], which suppresses regucalcin mRNA expression in the kidney cortex [38]. Proteases, which are activated by regucalcin, may be reduced in the kidney cortex cytoplasm of saline-administered rats. In addition, the saline administration-induced decrease in proteolytic activity of rat renal cortex cytoplasm is partly involved in the suppressed regucalcin expression.

The physiological role of regucalcin in the activation of thiol proteases in kidney cortex cells remains to be elucidated. However, regucalcin plays a role in the regulation of the degradation of proteins that are involved in the regulation of signalling pathways. In addition, regucalcin may play a part in the regulation of protein turnover in kidney cortex cells.

Regucalcin Regulates Nuclear Function

Nuclear Localisation of Regucalcin

Regucalcin has been found to localize in the nucleus of HA-regucalcin/phCMV2-transfected NRK52E cells using immunocytochemical analysis [54]. The nuclear localisation of regucalcin was enhanced after culture with BS, PTH, Bay K 8644 or PMA [54]. This enhancement was remarkable after culture with PMA. PMA-enhanced regucalcin expression is suppressed by staurosporine, an inhibitor of protein kinase C [54].

Regucalcin, which is enhanced through Ca^{2+}-signalling factors including protein kinase C, is localised in the nucleus of NRK52E cells. The results of Western blot analysis demonstrate that the nuclear regucalcin levels are markedly increased after culture with PMA [54]. This increase is depressed in the presence of staurosporine, an inhibitor of protein kinase C [80].

Thus, the results with Western blot analysis support the observation with immunocytochemical analysis for the nuclear localization of regucalcin in NRK52E cells. The nuclear localisation of regucalcin in stable regucalcin /pCXN2-transfected cells (transfectants) is markedly localized as compared with that of wild-type cells [54]. This increase is not seen in the transfectants cultured with staurosporine. These observations support the view that the localization of regucalcin into the nuclei is enhanced through Ca^{2+} signalling related to protein kinase C in NRK52E cells.

The expression of regucalcin mRNA in NRK52E cells may be partly mediated through signalling pathways that are related to cAMP or protein kinase C. The action of PTH is mediated through cAMP or IP_3-released Ca^{2+} and protein kinase C in cells [49].

The effect of PTH in stimulating the nuclear localisation of regucalcin in NRK52E cells may be mediated through cAMP and/or Ca^{2+}-dependent protein kinase C in NRK52E cells. In addition, PMA, an activator of protein kinase C, has been shown to enhance markedly the nuclear localisation of regucalcin in NRK52E cells [54].

Thus, protein kinase C may play a pivotal role in the enhancement of both regucalcin mRNA expression and the nuclear localisation of regucalcin protein in NRK52E cells. The nuclear localisation of regucalcin, which is enhanced through hormonal signalling, may play a role in the regulation of the expression of many genes in the nucleus.

Regucalcin Suppresses Nuclear DNA Synthesis

Regucalcin has been shown to have an inhibitory effect on DNA synthesis activity in the nuclei of normal rat liver [26]. Ca^{2+} is present in the liver nucleus. The nuclear DNA synthesis activity is increased in the presence of EGTA, a chelator of Ca^{2+}, in a reaction mixture *in* vitro [26], suggesting an inhibitory role of the nuclear endogenous Ca^{2+}. Regucalcin (0.1–0.5 μM) has a suppressive effect on DNA synthesis activity in the nuclei isolated from the rat renal cortex *in vitro* [81]. The effect of regucalcin is also seen in the presence of calcium chloride (50 μM) in the reaction mixture and is enhanced in the presence of EGTA (1 mM) [81].

Regucalcin may have an inhibitory effect on DNA synthesis in the nucleus of the renal cortex through a mechanism that is not related to Ca^{2+}. The presence of anti-regucalcin monoclonal antibody (10–50 ng/ml) in the reaction mixture caused a significant increase in the nuclear DNA synthesis activity [81]. This increase is completely abolished in the presence of regucalcin (0.5 μM). The nuclear endogenous regucalcin has been found to have a suppressive effect on DNA synthesis in the nuclei of the rat renal cortex [81].

The suppressive effect of regucalcin on nuclear DNA synthesis is also observed in the presence of various inhibitors (staurosporine, TFP or okadaic acid) for protein kinases and protein phosphatases, which are involved in the regulation of the nuclear functions including cell proliferation and gene expression, in the reaction mixture [81].

These results suggest that the effect of regucalcin in decreasing the nuclear DNA synthesis activity is not related to protein kinases and protein phosphatases in the renal cortex nucleus. Regucalcin may directly bind to DNA and inhibit DNA synthesis. Regucalcin may play a role as a suppressor in DNA synthesis activity in the renal cortex nucleus.

Role of Regucalcin in Proximal Tubular Epithelial Cells

Proximal tubular epithelial NRK52E cells are predominantly present in the kidney cortex. NRK52E cells are cloned from the normal rat kidney cortex. Regucalcin has been shown to express in proximal tubular epithelial NRK52E cells *in vitro* [44]. The role of regucalcin in the regulation of kidney cell function is examined using NRK52E cells.

Overexpression of Endogenous Regucalcin Suppresses Cell Proliferation

Regucalcin /pCXN2-transfected cells (transfectants) for NRK52E cells, which stably overexpress regucalcin, have been generated to determine the role of endogenous regucalcin in the regulation of cell function [82]. The regucalcin content of regucalcin /pCXN2-transfected cells was about 21-fold as compared with that of the parental wild-type NRK52E cells [82]. Overexpression of endogenous regucalcin has been found to have a role in the regulation of the proliferation of NRK52E cells [82]. Regucalcin may have a physiological role in the regulation of the proliferation of rat kidney proximal tubular epithelial cells. The effect of overexpression of regucalcin on cell proliferation is suppressed after culture with butyrate, roscovitine and sulforaphane, which induce cell cycle arrest [83]. Butyrate induces an inhibition of G1 progression [84]. Roscovitine is a potent and selective inhibitor of the cyclin-dependent kinase cdc2, cdk2 and cdck5 [85] and can arrest in G1 and accumulate in G2 phases of cell cycle. Sulforaphane can induce G2/M phase cell cycle arrest [86]. The effect of butyrate, roscovitin or sulforaphan, which inhibits the proliferation of wild-type NRK52E cells, is not seen in the transfectants overexpressing regucalcin [82]. This finding suggests that endogenous regucalcin induces G1 and G2/M phase cell cycle arrest in NRK52E cells. The proliferation of NRK52E cells is also suppressed after culture with PD98059, staurosporine or dibucaine, which is an inhibitor of various protein kinases [82]. These inhibitions are not seen in the transfectants that are overexpressing endogenous regucalcin [82]. The suppressive effect of endogenous regucalcin on cell proliferation may result from the inhibitory effect of regucalcin on various protein kinases that are involved in the stimulation of cell proliferation.

In addition, culture with wortmannin, an inhibitor of PI3-kinase [86], represses proliferation of NRK52E cells [82]. This repression is not observed in the transfectants [82]. Endogenous regucalcin may inhibit PI3-kinase and partly contribute to suppression of cell proliferation in NRK52E cells. The proliferation of wild-type NRK52E cells is repressed in the presence of Bay K 8644, an agonist of calcium entry into cells [82]. This effect is not seen in the transfectants [82]. This may support the view that endogenous regucalcin suppresses apoptotic cell death, which is mediated through an increase in intracellular Ca^{2+} levels due to maintaining intracellular calcium homeostasis.

Moreover, the gene expression for proteins that are involved in cell proliferation and cell cycle has been shown to regulate in the transfectants. The expression of c-jun and checkpoint-kinase 2 (chk2) mRNAs has been found to be suppressed in the transfectants [82]. The expression of p53 mRNA is enhanced in the transfectants, while the expression of c-myc, c-fos, cdc2 and p21mRNAs is not changed in the transfectants [82]. The decrease in c-jun and chk2 mRNA expressions may partly contribute to suppression of cell proliferation, which is induced in NRK52E cells overexpressing regucalcin. The expression of the tumour suppressor gene p53 mRNA in the transfectants may play a partial role in the retardation of proliferation of NRK52E cells. Regucalcin has been shown to localise in the nucleus of NRK52E cells [54]. Regucalcin may reveal a suppressive effect on cell proliferation due to regulating the gene expression of many proteins that are related to cell proliferation in NRK52E cells.

Thus, overexpression of endogenous regucalcin has been demonstrated to reveal a suppressive effect on the cell proliferation due to inducing G1 and G2/M phase cell cycle arrest in NRK52E cells; the effect of regucalcin may be mediated through a decrease in various Ca^{2+} signalling-dependent protein kinases and PI3-kinase activities and suppression of c-jun and chk2 mRNAs expression or the enhancement of p53 mRNA expression.

Overexpression of Endogenous Regucalcin Suppresses Apoptotic Cell Death

The role of endogenous regucalcin in apoptotic cell death is shown in NRK52E cells overexpressing regucalcin [87]. The number of wild-type cells was decreased after culture for 42–72 hours in the presence of TNF-α (0.1 or 1.0 ng/ml of medium), lypopolysaccharide (LPS; 0.1 or 1.0 μg/ml), Bay K 8644 (1–100 nM), or thapsigargin (1–100 nM) [87].

These effects are not seen in the transfectants that overexpress regucalcin. DNA fragmentation was induced after culture with LPS, Bay K 8644 or thapsigargin, and these effects are depressed in the transfectants [87]. Overexpression of regucalcin has been found to have a suppressive effect on apoptotic cell death induced by TNF-α, LPS, Bay K 8644 or thapsigargin in NRK52E cells [87].

TNF-α, LPS or Bay K 8644-induced cell death in NRK52E cells is repressed in the presence of caspase-3 inhibitor [87]. LPS or Bay K 8644-induced cell death is blocked by N ω-nitro-L-arginine methylester (NAME), an inhibitor of NO synthase, in NRK52E cells [87]. Thapsigargin-induced cell death is repressed in the presence of caspase-3 inhibitor or NAME. NO may be important as a signalling factor in many cell types [73] and plays a role in apotosis of hepatoma cells [88]. NO is produced by NO synthase that is activated by Ca^{2+}/calmodulin [73]. Bay K 8644-induced calcium entry into the cells induces cell death [89]. The effect of regucalcin in suppressing apoptotic cell death may be mediated through its action on many intracellular signalling pathways in NRK52E cells.

Bcl-2 is a suppressor in apoptotic cell death [90]. Apaf-1 participates in activation of caspase-3 [91]. Akt-1 is involved in survival signalling pathways for cell death [92]. Overexpression of regucalcin has been found to produce a remarkable elevation of Bcl-2 mRNA expression in NRK52E cells and slightly stimulated Akt-1 mRNA expression in the cells [87]. Apaf-1, caspase-3 or G3PDH mRNA expressions are not altered in the transfectants [87]. The enhancement of Bcl-2 mRNA expression may contribute to the suppression of apoptotic cell death in NRK52E cells overexpressing regucalcin. Endogenous regucalcin may play a role in the regulation of Bcl-2 gene expression in NRK52E cells. The expression of caspase-3 mRNA in NRK52E cells is enhanced after culture with TNF-α [87]. This enhancement is suppressed in the transfectants [87]. The mechanism by which regucalcin suppresses TNF-α-induced cell death may be partly related to the decrease in caspase-3 mRNA expression in the transfectants.

Culture with LPS caused a decrease in Bcl-2 mRNA expression in NRK52E cells, suggesting that this decrease leads to LPS-induced cell death [87]. Overexpression of endogenous RGN enhances Bcl-2 mRNA expression, and this effect is also seen in the presence of LPS [87]. LPS-stimulated Apaf-1 mRNA expression is suppressed in the transfectants [91]. This may lead to suppression of LPS-induced cell death in NRK52E cells. Culture with Bay K 8644 or thapsigargin has been found to produce an increase in caspase-3 mRNA expression in wild-type NRK52E cells, suggesting that the increase in the gene expression induces apoptotic cell death [87]. This increase is depressed in the transfectants [87]. Regucalcin may have a suppressive effect on caspase-3 mRNA expression enhanced by Bay K 8644 or thapsigargin in NRK52E cells. Thus, overexpression of endogenous regucalcin has been found to enhance the expression of Bcl-2 and Akt-1 mRNAs and to suppress the expression of caspase-3 and Apaf-1 mRNAs in NRK52E cells.

Many toxic factors have been reported to induce renal failure due to stimulating apoptotic cell death [92]. Overexpression of regucalcin has a suppressive effect on apoptotic cell death induced by various factors (including TNF-α, LPS, Bay K 8644 or thapsigargin) in NRK52E cells [87]. Endogenous regucalcin may play an important role as a suppressor in the development of apoptotic cell death in the kidney proximal tubular epithelial cells.

Overexpression of Endogenous Regucalcin Regulates Gene Expression of Mineral Ion Transport-Related Proteins

Overexpression of endogenous regucalcin has been shown to increase in rat outer medullary K^+ channel (ROMK) mRNA expression in NRK52E cells, while it did not have an effect on Na^+, K^+-ATPase and epithelial sodium channel (ENaC) mRNA expressions [93]. Culture with aldosterone causes an increase in ENaC, Na^+, K^+-ATPase and ROMK mRNA expression in wild-type NRK52E cells [93]. The expression of these genes has been reported to increase after treatment with aldosterone in rat kidney cells [94-97]. The effect of aldosterone in increasing ENaC and Na^+, K^+-ATPase mRNA expressions is not observed in the transfectants overexpressing regucalcin [93]. However, overexpression of endogenous regucalcin reveals a stimulatory effect on ROMK mRNA expression in the transfectants untreated with aldosterone [93]. Culture with aldosterone does not enhance ROMK mRNA expression in the transfectants cultured with serum withdrawal [93]. Aldosterone has been shown to up-regulate regucalcin mRNA expression in NRK52E cells [44]. The stimulatory effect of aldosterone on ENaC and Na^+, K^+-ATPase mRNA expressions may be partly mediated through endogenous regucalcin in NRK52E cells. The expression of type II Na-Pi cotransporter (NaPi-IIa) and angiotensinogen mRNAs is not changed in NRK52E cells overexpressing regucalcin [93], suggesting that endogenous regucalcin does not have effects on NaPi-IIa and angiotensinogen mRNA expressions in NRK52E cells.

Overexpression of endogenous regucalcin has also been found to have suppressive effects on the gene expression of L-type Ca^{2+} channel and calcium-sensing receptor (CaR), which regulate intracellular Ca^{2+} signalling in NRK52E cells [93]. The blockade of Ca^{2+} influx through L-type Ca^{2+} channels has been shown to attenuate mitochondrial injury and apoptosis in hypoxia of renal tubular cells [98].

The entry of Ca^{2+} through L-type Ca^{2+} channels induces mitochondrial disruption and cell death [98]. CaR participates in the regulation of renal Ca^{2+} transport [99, 100]. Regucalcin may regulate the intracellular Ca^{2+}-signalling pathway, which is mediated through its suppressive effect on L-type Ca^{2+} channel or CaR mRNA expressions in the kidney proximal tubular epithelial cells.

The expression of regucalcin mRNA in NRK52E cells has been shown to enhance after PTH treatment [44], suggesting that regucalcin partly mediates cellular response for PTH in kidney cells. Overexpression of endogenous regucalcin did not attenuate the expression of L-type Ca^{2+} channel or CaR mRNAs, which is decreased after PTH treatment, in NRK52E cells [93]. Endogenous regucalcin is found to suppress L-type Ca^{2+} channel or CaR mRNA expressions in NRK52E cells [93].

Regucalcin may play a role as a mediator in the cellular response after stimulation with PTH in NRK52E cells.

Regucalcin has been shown to play a role in the regulation of intracellular Ca^{2+} transport; the protein activates Ca^{2+}-pumping enzymes (Ca^{2+}-ATPase) in the basolateral membranes [46], mitochondria [61] and microsomes [60] in the rat kidney cortex. Regucalcin may regulate intracellular Ca^{2+} homeostasis in the kidney proximal tubular epithelial cells that Ca^{2+} is passed through transcellular transport.

Moreover, regucalcin has been shown to suppress the expression of L-type Ca^{2+} channel or CaR mRNAs in NRK52E cells [93].

Interestingly, calcium has been demonstrated to be present through numerous steps of tubulogenesis and nephron induction during embryonic development of the kidney [94]. Several calcium-binding proteins such as

RGN and calbindin-D28k are commonly used to label pronephric tubules and metanephric ureteral epithelium [94].

Regucalcin may play a physiological role in the regulation of intracellular Ca^{2+} homeostasis, mineral transport, cell proliferation and apoptotic cell death in the kidney proximal tubular epithelial cells as shown in Figure 1. Regucalcin, which is translocated into the nucleus through the Ca^{2+} signalling pathway, regulates the gene expression of their many related proteins.

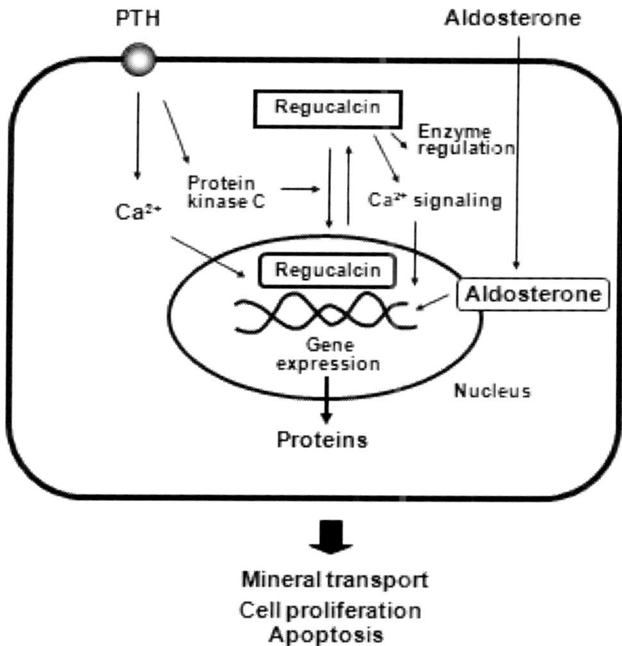

Figure 1. Regucalcin plays a multifunctional role in the kidney proximal tubular epithelial cells. Endogenous regucalcin plays a role in the regulation of mineral transport, cell proliferation and apoptotic cell death in the kidney proximal tubular epithelial cells. The expression of regucalcin gene is stimulated through Ca^{2+} signalling, parathyroid hormone or aldosterone, which regulates mineral transport in the cells. Regucalcin suppresses many enzyme activations, which are mediated through Ca^{2+} and other signalling pathways. Nuclear localisation of regucalcin is promoted through protein kinase C-related signalling. Nuclear RGN regulates the gene expression of many proteins that mediate mineral transport, cell proliferation and apoptotic cell death.

Overexpression of Endogenous Regucalcin Suppresses TNF-α- and TGF-β1-Mediated Cell Responses

Overexpression of endogenous regucalcin, moreover, has been found to have a suppressive effect on cell responses that are mediated through the signalling process following stimulation with TNF-α or TGF-β1 in NRK52E cells [55]. Regucalcin mRNA expression is suppressed after TNF-α or TGF-β1 in NRK52E cells [44]. However, overexpression of endogenous regucalcin suppressed apoptotic cell death induced by TNF-α or TGF-β1 that is mediated through caspase-3 in NRK52E cells [55].

Regucalcin localizes in the nucleus of NRK52E cells [54] and may inhibit nuclear DNA fragmentation that is partly related to caspase-3.

The effect of Ca^{2+}/calmodulin in increasing NO synthase activity in NRK52E cells is depressed in the transfectants overexpressing regucalcin [55]. NO synthase activity is increased in wild-type NRK52E cells after culture with TNF-α, although culture with TGF-β1 does not have an effect on the enzyme activity [55]. Overexpression of endogenous regucalcin has a suppressive effect on the increase in NO synthase activity in NRK52E cells cultured with TNF-α [55]. The suppressive effect of regucalcin on TNF-α-induced cell death may be partly involved in its inhibitory effect on NO synthase activity in NRK52E cells.

Culture with TNF-α or TGF-β1 causes a remarkable increase in α-smooth muscle actin level in NRK52E cells [55]. Interestingly, this increase is not seen in the transfectants. In addition, the expression of α-smooth muscle actin is markedly suppressed in the transfectants cultured without TNF-α or TGF-β1 [55]. These findings suggest that overexpression of endogenous regucalcin has a suppressive effect on the expression of α-smooth muscle actin in NRK52E cells cultured with TNF-α or TGF-β1 and that regucalcin regulates the signalling pathway that is mediated through TNF-α or TGF-β1. TGF-β1 is a key mediator that regulates transdifferentiation of NRK52E cells into myofibroblasts expressing α-smooth muscle actin [101]. This may lead to renal fibrosis associated with overexpression of TGF-β1 within the diseased kidney [101]. Regucalcin may suppress transdifferentiation to renal fibrosis in NRK52E cells with TGF-β1 or TNF-α.

Overexpression of endogenous regucalcin causes a significant increase in the expression of mRNAs of Smad 2, which is involved in signal transduction of TGF-β1 [102], or NF-κB, which is related to signalling of TNF-α [103], in NRK52E cells [55]. Culture with TGF-β1 or TNF-α causes a remarkable increase in the expression of mRNAs of Smad 2 or NF-κB in wild-type

NRK52E cells, respectively [55]. However, these cytokines do not enhance the gene expression of Smad 2 and NF-κB in the transfectants [55]. Regucalcin may stimulate the gene expression of Smad 2 or NF-κB, which is related to signalling pathways of TNF-α or TGF-β1. In addition, overexpression of endogenous regucalcin may have a suppressive effect on the signalling pathway by which TNF-α or TGF-β1 stimulates the gene expression of NF-κB or Smad 2 in NRK52E cells. The suppressive effects of regucalcin on inducing apoptotic cell death and expression of α-smooth muscle actin may not be mediated through the expression of NF-κB or Smad 2 that is stimulated by TNF-α or TGF-β1 in NRK52E cells.

Thus, overexpression of endogenous regucalcin has a suppressive effect on cell responses that are mediated through intracellular signalling pathways for stimulation of TNF-α or TGF-β1 in NRK52E cells.

Involvement of Regucalcin in Kidney Disorder

There is growing evidence that the suppression of regucalcin gene expression is involved in the development of kidney disorder. The tubular epithelial cells in regucalcin gene knockout mice aged 12 months have been shown to cause deposition of lipofuscin and presence of senescence-associated beta-galactosidase [104]. Regucalcin mRNA expression is suppressed in the kidney cortex of rats after saline intake for 7 days, which causes an increase in the serum calcium and BUN concentrations as an index of kidney disorder [38-40]. Regucalcin mRNA expression is suppressed in SHR, suggesting an involvement in a hypertensive state [38].

The expression of regucalcin mRNA is suppressed in the renal cortex of rats with chemically induced kidney damage [43]. Regucalcin mRNA expression is markedly reduced in the kidney cortex of rats that received a single intraperitoneal administration of cisplatin and cephaloridine; these administrations induced a remarkable accumulation of calcium in the kidney cortex and a corresponding elevation of BUN at 1–3 days after administration [43]. Kidney toxic chemicals-induced suppression of regucalcin expression may lead to renal disease.

The expression of regucalcin gene has been shown to be down-regulated in rat kidney tissues after treatment with many nephrotoxicants [105]. The proximal tubule segment-specific nephrotoxicants, namely hexachloro-1:3-

butadiene (HCBD), specific for S(3) segment (pars recta), and potassium dichromate (chromate) specific for S(1)–S(2) segments (pars convoluta) were used in this study [105]. Regucalcin is down-regulated after HCBD administration with low dose, whereas chromate causes the same effect at high doses only [105]. Glutamine synthase activity in the kidney cortex shows a similar behaviour, even if sensitive to low doses of chromate also, whereas BUN and creatinine increase after a high dose of both chemicals only [105]. The expression of regucalcin gene appears to be a sensitive genomic marker to evaluate the renal impairment caused by chemicals and its down-regulation seems to be related to damage, early or already established, to the S(3) segment of the proximal tubule [105].

Ochratoxin A (OTA), a naturally occurring mycotoxin, is nephrotoxic in all animal species tested and is considered a potent renal carcinogen, although its mechanism of toxicity is still unknown [106]. Rats were gavaged daily with OTA (0.5 mg/kg body weight), and gene expression profiles in target and non-target organs were analyzed after administration for 7 and 21 days [106]. The number of differentially expressed genes in the kidney was much higher than those in the liver (541 versus 11 at both time points) [106]. Down-regulation of genes was the predominant effect, and the oxidative stress response pathway and genes involved in metabolism and transport were inhibited at both time points [106].

Regucalcin was strongly inhibited at both time points, and the genes implicated in cell survival and proliferation were up-regulated at day 21 [106]. Moreover, the translation factors and annexin genes were up-regulated at both time points [106]. Apart from oxidative stress, alterations of the calcium homeostasis and cytoskeleton structure may be present at the first events of OTA toxicity [106].

Prolonged intake of aristolochic acid (AA) has been shown to be associated with development of certain renal disorders [107]. Renal tubular atrophy and interstitial fibrosis are the early symptoms of AA nephropathy. Symptoms are observed in rats that are dosed with AA at a dosage of 10 mg/kg body/day for 1 month [107]. Differentiated proteins are identified in the kidney tissues from proteomics investigations [107]. The up-regulated proteins identified included ornithine aminotransferase, sorbitol dehydrogenase, actin, aspartoacylase, 3-hydroxyisobutyrate dehydrogenase and peroxiredoxin-1 [107]. The down-regulated proteins included regucalcin, ATP synthase subunit β, glutamate dehydrogenase 1, glutamate-cysteine ligase regulatory subunit, dihydropteridine reductase, hydroxyacyl-coenzyme A dehydrogenase, voltage-dependent anion-selective channel protein 1, prohibitin and adenylate kinase isoenzyme 4 [107]. Several identified protein markers were found to have a biological and medical significance.

The basic mechanism underlying calcineurin inhibitor (cyclosporine) nephrotoxicity with enhancement by sirolimus is still largely unknown [108]. The effects of cyclosporine (10 mg/kg body weight/day) alone and in combination with sirolimus (1 mg/kg/day) on the renal proteome and the correlation of these effects with urine metabolite pattern changes have been investigated [108]. After 28 days, 24-hour urine was collected for NMR-based metabolic analysis, and the kidneys were harvested for two dimensional-gel electrophoresis and histology.

Cyclosporine affected the following groups of proteins: calcium homeostasis (regucalcin and calbindin), cytoskeleton (vimentin and caldesmon), response to hypoxia and mitochondrial function (prolyl 4-hydroxylase, proteasome and NADH dehydrogenase) and cell metabolism (kidney aminoacylase, pyruvate dehydrogenase and fructose-1, 6-bisphosphate) [108]. Regucalcin increased in the urine with kidney disorder and may be useful as a biomarker in kidney disorders.

As described above, kidney regucalcin expression has been demonstrated to be down-regulated with renal disorder induced by various drugs. Depression of the regucalcin gene expression in kidney cells may play a pathophysiological role in development of renal disorders.

Conclusion

Calcium, which plays a multifunctional role in many biological systems, plays a pivotal role during embryonic development of kidneys and is present through numerous steps of tubulogenesis and nephron induction from the formation of a simple kidney in amphibian larvae, the pronephrone, to the formation of the more complex mammalian kidney, the metanephrons. Regucalcin, which is a Ca^{2+}-binding protein [4-7], is greatly expressed in proximal tubular cells. Several Ca^{2+}-binding proteins such as regucalcin and calbindin-D28k are commonly used to label pronephric tubules and metanephric ureteral epithelium. Regucalcin is focused on Ca^{2+}-binding proteins and Ca^{2+} sensors that are involved in renal organogenesis in the link between Ca^{2+}-dependent signals and polycystins. Regucalcin may be involved in tubulogenesis and nephron induction. Regucalcin plays a physiological role in the regulation of Ca^{2+} homeostasis due to activating membranous Ca^{2+}-

pump enzymes in the kidney proximal tubular epithelial cells, which participate in transcellular transport to reabsorption of calcium from filtrated urinary calcium. Regucalcin suppresses the activities of various Ca^{2+}-dependent protein kinases and protein phosphatases, NO synthase and cAMP phosphodiesterase, which are related to intracellular signalling pathways. Moreover, regucalcin suppresses nuclear DNA synthesis in the kidney cortex and activates thiol proteases. Thus, regucalcin plays a pivotal role in the regulation of kidney cell function. Further findings of the role of regucalcin in kidney cell regulation are expected. Regucalcin localises to the nucleus, and it has a regulatory effect on the gene expression of various proteins in the nucleus of kidney tubular epithelial cells.

Regucalcin regulates the gene expressions of mineral transport-related proteins, Smad 2 and NF-κB, and cell proliferation and apoptosis-related proteins. Interestingly, regucalcin suppresses the gene expression of TNF-α and TGF-β1-stimulated α-smooth muscle actin that induces trans-differentiation to renal fibrosis, suggesting an involvement in the development of renal disease. Future studies will determine the molecular mechanism by which regucalcin regulates gene expression in the nucleus of kidney tubular epithelial cells.

The expression of regucalcin gene is suppressed in various patho-physiological states with hypertensive state or drugs that induce kidney disorder. The analysis for proteome and differential gene expression shows a unique suppression of regucalcin expression among many proteins. The suppression of the regucalcin gene expression is associated with the development of kidney disorder, suggesting a pathophysiological role.

The role of regucalcin in the kidney function of humans is not still determined. This remains to be elucidated in future studies. Targeting for the regucalcin gene may help in the prevention and restoration of kidney disorder. Clinical studies for regucalcin are expected.

References

[1]   Cheung, W. Y. (1980) Calmodulin plays a pivotal role in cellular regulation. *Science* 202:19–27.

[2]   Nishizuka, Y. (1986) Studies and perspectives of protein kinase C. *Science* 233:305–312.

[3] Kraus-Friedman, N., Feng, L. (1996) The role of intracellular Ca^{2+} in the regulation of gluconeogenesis. *Metabolism* 45:389–403.

[4] Yamaguchi, M., Yamamoto, T. (1978) Purification of calcium binding substance from soluble fraction of normal rat liver. *Chem. Pharm. Bull.* 26:1915–1918.

[5] Yamaguchi, M., Yoshida, H. (1985) Regulatory effect of calcium-binding protein isolated from rat liver cytosol on activation of fructose 1,6-diphosphatase by Ca^{2+}-calmodulin. *Chem. Pharm. Bull.* 33:4489–4493.

[6] Yamaguchi, M., Mori, S. (1988) Effect of Ca^{2+} and Zn^{2+} on 5'-nucleotidase activity in rat liver plasma membranes: hepatic calcium-binding protein (regucalcin) reverses the Ca^{2+} effect. *Chem. Pharm. Bull.* 36:321–325.

[7] Yamaguchi, M. (1992) A novel Ca^{2+}-binding protein regucalcin and calcium inhibition. Regulatory role in liver cell function. In: Kohama, K., editor. *Calcium inhibition.* Tokyo: Japan Sci Soc Press, p.19–41.

[8] Yamaguchi, M. (2011) The transcriptional regulation of regucalcin gene expression. *Mol. Cell. Biochem.* 346:147–171.

[9] Shimokawa, N., Yamaguchi, M. (1993) Molecular cloning and sequencing of the cDNA coding for a calcium-binding protein regucalcin from rat liver. *FEBS Lett.* 327:251–255.

[10] Misawa, H., Yamaguchi, M. (2000) The gene of Ca^{2+}-binding protein regucalcin is highly conserved in vertebrate species. *Int. J. Mol. Med.* 6: 191–196.

[11] Shimokawa, N., Matsuda, Y., Yamaguchi, M. (1995) Genomic cloning and chromosomal assignment of rat regucalcin gene. *Mol. Cell. Biochem.* 151:157–163.

[12] Thiselton, D. L., McDowall, J., Brandau, O., Ramser, J., d'Esposito, F., Bhattacharga, S. S., et al. (2002) An integrated, functionally annotated gene map of the DXS8026-ELK1 internal on human Xp11.3-Xp11.23: potential hotspot for neurogenetic disorders. *Genomics* 79:560–572.

[13] Yamaguchi, M., Makino, R., Shimokawa, N. (1996) The 5'end sequences and exon organization in rat regucalcin gene. *Mol. Cell. Biochem.* 165:145–150.

[14] Shimokawa, N., Yamaguchi, M. (1992) Calcium administration stimulates the expression of calcium-binding protein regucalcin mRNA in rat liver. *FEBS Lett.* 305:151–154.

[15] Murata, T., Yamaguchi, M. (1998) Ca^{2+} administration stimulates the binding of AP-1 factor to the 5'-flanking region of the rat gene for the Ca^{2+}-binding protein regucalcin. *Biochem. J.* 329:157–163.

[16] Murata, T., Yamaguchi, M. (1999) Promoter characterization of the rat gene for Ca^{2+}-binding protein regucalcin. Transcriptional regulation by signaling factors. *J. Biol. Chem.* 274:1277–1285.

[17] Misawa, H., Yamaguchi, M. (2002) Indentification of transcription factor in the promoter region of rat regucalcin gene: binding of nuclear factor I-A1 to TTGGC motif. *J. Cell. Biochem.* 84:795–802.

[18] Yamaguchi, M. (2009) Novel protein RGPR-p117: its role as the regucalcin gene transcription factor. *Mol. Cell. Biochem.* 327:53–63.

[19] Nejak-Bowen, K. N., Zeng, G., Tan, X., Cieply, B., Monga, S. P. (2009) β-Catenin regulates vitamic C biosynthesis and cell survival in murine liver. *J. Biol. Chem.* 284:28115–127.

[20] Yamaguchi, M., Isogai, M. (1993) Tissue concentration of calcium-binding protein regucalcin in rats by enzyme-linked immunoadsorbent assay. *Mol. Cell. Biochem.* 122:65–68.

[21] Yamaguchi, M. (2005) Role of regucalcin in maintaining cell homeostasis and function (review). *Int. J. Mol. Med.* 15:371–389.

[22] Yamaguchi, M. (2011) Regucalcin and cell regulation: role as a supressor in cell signaling. *Mol. Cell. Biochem.* 353:101–137.

[23] Laurentino, S. S., Correia, S., Cavaco, J. E., Oliveira, P. F., de Sousa, M., Barros, A., et al. (2012) Regucalcin, a calcium-binding protein with a role in male reproduction? *Mol. Hum. Reprod.* 18:161–170.

[24] Takahashi, H., Yamaguchi, M. (1997) Stimulatory effect of regucalcin on ATP-dependent calcium transport in rat liver plasma membranes. *Mol. Cell. Biochem.* 168:149–153.

[25] Tsurusaki, Y., Yamaguchi, M. (2004) Role of regucalcin in liver nuclear function: binding of regucalcin to nuclear protein or DNA and modulation of tumor-related gene expression. *Int. J. Mol. Med.* 14:277–281.

[26] Tsurusaki, Y., Yamaguchi, M. (2002) Suppressive role of endogenous regucalcin in the enhancement of deoxyribonucleic acid synthesis activity in the nucleus of regenerating rat liver. *J. Cell. Biochem.* 85:516–522.

[27] Tsurusaki, Y., Yamaguchi, M. (2002) Role of endogenous regucalcin in nuclear regulation of regenerating rat liver: suppression of the enhanced ribonucleic acid synthesis activity. *J. Cell. Biochem.* 87:450–457.

[28] Inagaki, S., Yamaguchi, M. (2001) Suppressive role of endogenous regucalcin in the enhancement of protein kinase activity with proliferation of cloned rat hepatoma cells (H4-II-E). *J. Cell. Biochem.* Suppl. 36:12–18.

[29] Izumi, T., Yamaguchi, M. (2004) Overexpression of regucalcin suppresses cell death in cloned rat hepatoma H4-II-E cells induced by tumor necrosis factor-α or thapsigargin. *J. Cell. Biochem.* 92:296–306.

[30] Fujita, T., Uchida, K., Maruyama, N. (1992) Purification of senescence marker protein-30 (SMP30) and its androgen-independent decrease with age in the rat liver. *Biochim. Biophys. Acta* 1116:122–128.

[31] Fujita, T., Shirasawa, T., Uchida, K., Maruyama, N. (1992) Isolation of cDNA clone encoding rat senescence marker protein-30 (SMP30) and its tissue distribution. *Biochim. Biophys. Acta* 1132:297–305.

[32] Yamaguchi, M., Kurota, H. (1995) Expression of calcium-binding protein regucalcin mRNA in the kidney cortex of rats: the stimulation by calcium administration. *Mol. Cell. Biochem.* 146:71–77.

[33] Van Os, C. H. (1987) Transcellular calcium transport in intestinal and renal epithelial cells. *Biochim. Biophys. Acta* 906:195–222.

[34] Taylor, C. W. (1985) Calcium regulation in vertebrates: an overview. *Comp. Biochem. Physiol.* 82A:249–255.

[35] Murata, T., Yamaguchi, M. (1999) Binding of kidney nuclear proteins to the 5' flanking region of the rat gene for Ca^{2+}-binding protein regucalcin: involvement of Ca^{2+}/calmodulin signaling. *Mol. Cell. Biochem.* 199:35–40.

[36] Misawa, H., Yamaguchi, M. (2001) Involvement of nuclear factor-1 (NFl) binding motif in the regucalcin gene expression of rat kidney cortex: the expression is suppressed by cisplatin administration. *Mol. Cell. Biochem.* 219:29–37.

[37] Kurota, H., Yamaguchi, M. (1996) Steroid hormonal regulation of calcium-binding protein regucalcin mRNA expression in the kidney cortex of rats. *Mol. Cell. Biochem.* 155:105–111.

[38] Shinya, N., Kurota, H., Yamaguchi, M. (1996) Calcium-binding protein regucalcin mRNA expression in the kidney cortex is suppressed by saline ingestion in rats. *Mol. Cell. Biochem.* 162:139–144.

[39] Shinya, N., Yamaguchi, M. (1997) Alteration in Ca^{2+}-ATPase activity and calcium-binding protein regucalcin mRNA expression in the kidney cortex of rats with saline ingestion. *Mol. Cell. Biochem.* 170:17–22.

[40] Shinya, N., Yamaguchi, M. (1998) Stimulatory effect of calcium administration on regucalcin mRNA expression is attenuated in the

kidney cortex of rats ingested with saline. *Mol. Cell. Biochem.* 178:275–281.

[41] Montine, T. J., Borch, R. F. (1990 Role of endogenous sulfur-containing nucleotides in an *in vitro* model of cis-dismminechloro platinum (II)-induced nephrotoxicity. *Biochem. Phramacol.* 39:1751–1757.

[42] Goldstein, R. S., Pasino, D. A., Hewitt, W. R., Hook, J. B. (1986) Biochemical mechanism of cepharoridine nephrotoxicity: time and concentration dependence of peroxidative injury. *Toxicol. Appl. Pharmacol.* 83:261–270.

[43] Kurota, H., Yamaguchi, M. (1995) Suppressed expression of calcium binding protein regucalcin mRNA in the renal cortex of rats with chemically induced kidney damage. *Mol. Cell. Biochem.* 151:55–60.

[44] Nakagawa, T., Yamaguchi, M. (2005) Hormonal regulation on regucalcin mRNA expression in cloned normal rat kidney proximal tubular epithelial NRK52E cells. *J. Cell. Biochem.* 95:589–597.

[45] Agus, Z. S., Chiu, P. J. S., Goldberg, M. (1997) Regulation of urinary calcium-excretion in the rat. *Am. J. Physiol.* 232:F545–F549.

[46] Kurota, H., Yamaguchi, M. (1997) Activatory effect of calcium-binding protein regucalcin on ATP-dependent calcium transport in the basolateral membranes of rat kidney cortex. *Mol. Cell. Biochem.* 169: 149–156.

[47] Van Heeswijk, M. P. E., Geertsen, J. A. M., Van Os, C. H. (1984) Kinetic properties of the ATP-dependent Ca^{2+} pump and the Na^{+}/Ca^{2+} exchange system in basolateral membranes from rat kidney cortex. *J. Membrane Biol.* 79:19–31.

[48] Rasmussen, J. (1970) Cell communication, calcium ion, and cyclic adenosine monophosphate. *Science* 170:404–412.

[49] Verheijem, M. H. G., Defize, L. H. K. (1997) Parathyroid hormone activates mitogen-activated protein kinase via a cAMP-mediated pathway independent of rats. *J. Biol. Chem.* 272:3423–3429.

[50] Vincenzi, F. F. (1982) Pharmacology of calmodulin antagonism. In: Godfraind, T., Albertini, A., Paoletti, R., editors. *Calcium modulators.* Amsterdam: Elsevier Biomedical Press, p.67–80.

[51] Liu, Y., Yang, Y., Ye, Y. C., Shi, Q. F., Chai, K., Tashiro, S., et al. (2012) Activation of ERK-p53 and ERK-mediated phosphorylation of Bcl-2 are involved in autophagic cell death induced by the c-Met inhibitor SU11274 in human lung cancer A549 cells. *J. Phramacol. Sci.* 118: 423–432.

[52] Chen, Q. W., Edvinsson, L., Xu, C. B. (2009) Role of ERK/MAPK in endothelin receptor signaling in human aortic smooth muscle cells. *BMC Cell Biol.* 10:52–64.

[53] Nakagawa, T., Yamaguchi, M. (2008) Nuclear localization of regucalcin is enhanced in culture with protein kinase C activation in cloned normal rat kidney proximal tubular epithelial NRK52E cells. *Int. J. Mol. Med.* 21: 605–610.

[54] Hunter, T. (1995) Protein kinases and phosphatases: the Yin and Yang of protein phosphorylation and signaling. *Cell* 80:225–236.

[55] Nakagawa, T., Yamaguchi, M. (2007) Overexpression of regucalcin suppresses cell response for tumor necrosis factor-α or transforming growth factor-β1 in cloned normal rat kidney proximal tubular epithelial NRK52E cells. *J. Cell. Biochem.* 100:1178–1190.

[56] Sawada, N., Yamaguchi, M. (2005) A novel regucalcin gene promoter region-related protein: comparison of nucleotide and amino acid sequences in vertebrate species. *Int. J. Mol. Med.* 15:97–104.

[57] Sawada, N., Yamaguchi, M. (2005) Overexpression of RGPR-p117 enhances regucalcin gene expression in cloned normal rat kidney proximal tubular epithelial cells. *Int. J. Mol. Med.* 16:1049–1055.

[58] Sawada, N., Yamaguchi, M. (2006) Overexpression of RGPR-p117 enhances regucalcin gene promoter activity in cloned normal rat kidney proximal tubular epithelial cells: involvement of TTGGC motif. *J. Cell. Biochem.* 99:589–597.

[59] Sawada, N., Yamaguchi, M. (2006) Involvement of nuclear factor I-A1 in the regulation of regucalcin gene promoter activity in cloned normal rat kidney proximal tubular epithelial cells. *Int. J. Mol. Med.* 18:665–671.

[60] Kurota, H., Yamaguchi, M. (1997) Regucalcin increases Ca^{2+}-ATPase activity and ATP-dependent calcium uptake in the microsomes of rat kidney cortex. *Mol. Cell. Biochem.* 177:201–207.

[61] Xue, J. H., Takahashi, H., Yamaguchi, M. (2000) Stimulatory effect of regucalcin on mitochondrial ATP-dependent calcium uptake activity in rat kidney cortex. *J. Cell. Biochem.* 80:285–292.

[62] Kennedy, M. B., Bennett, M. K., Erondu, N. E., Miller, S. G. (1987) Calcium/ calmodulin-dependent protein kinases. In: Cheung, W. Y., editor. *Calcium and cell function.* New York: Acadamic Press Inc, p.61–107.

[63] Magno, A. L., Ward, B. K., Batajczak, T. (2011) The calcium-sensing receptor: a molecular perspective. *Endocr. Rev.* 32:3–30.

[64] Kurota, H., Yamaguchi, M. (1997) Inhibitory effect of regucalcin on Ca^{2+}/calmodulin-dependent protein kinase activity in rat renal cortex cytosol. *Mol. Cell. Biochem.* 177:239–243.

[65] Kurota, H., Yamaguchi, M. (1998) Inhibitory effect of calcium binding protein regucalcin on protein kinase C activity in rat renal cortex cytosol. *Biol. Pharm. Bull.* 21:315–318.

[66] Pallen, C. J., Wang, J. H. (1983) Calmodulin-stimulated dephosphory-lation of *p*-nitrophenylphosphate and free phosphotyrosine by calcineurin. *J. Biol. Chem.* 258:8550–8553.

[67] Omura, M., Kurota, H., Yamaguchi, M. (1998) Inhibitory effect of regucalcin on Ca^{2+}/calmodulin-dependent phosphatase activity in rat renal cortex cytosol. *Biol. Pharm. Bull.* 21:440–443.

[68] Omura, M., Yamaguchi, M. (1998) Inhibition of Ca^{2+}/calmodulin-dependent phosphatase activity by regucalcin in rat liver cytosol: involvement of calmodulin binding. *J. Cell. Biochem.* 71:140–148.

[69] Morooka, Y., Yamaguchi, M. (2001) Suppressive role of endogenous regucalcin in the regulation of protein phosphatase activity in rat renal cortex cytosol. *J. Cell. Biochem.* 81:639–646.

[70] Morooka, Y., Yamaguchi, M. (2001) Inhibitory effect of regucalcin on protein phosphatase activity in the nuclei of rat kidney cortex. *J. Cell. Biochem.* 83:111–120.

[71] Morooka, Y., Yamaguchi, M. (2002) Endogenous regucalcin suppresses the enhancement of protein phosphatase activity in the cytosol and nucleus of kidney cortex in calcium-administered rats. *J. Cell. Biochem.* 85: 553–560.

[72] Yamaguchi, M., Kurota, H. (1997) Inhibitory effect of regucalcin on Ca^{2+}/calmodulin-dependent cyclic AMP phosphodiesterase activity in rat kidney cytosol. *Mol. Cell. Biochem.* 177:209–214.

[73] Lowenstein, C. J., Dinerman, J. L., Snyder, S. H. (1994) Nitric oxide: A physiologic messenger. *Ann. Intern. Med.* 120:227–237.

[74] Yamaguchi, M., Takahashi, H., Tsurusaki, Y. (2003) Suppressive role of endogenous regucalcin in the enhancement of nitric oxide synthase activity in liver cytosol of normal and regucalcin transgenic rats. *J. Cell. Biochem.* 88:1226–1234.

[75] Ma, Z. J., Yamaguchi, M. (2003) Regulatory effect of regucalcin on nitric oxide synthase activity in rat kidney cortex cytosol: role of endogenous regucalcin in transgenic rats. *Int. J. Mol. Med.* 12:201–206.

[76] Baba, T., Yamaguchi, M. (1999) Stimulatory effect of regucalcin on proteolytic activity in rat renal cortex cytosol: involvement of thiol proteases. *Mol. Cell. Biochem.* 195:87–92.

[77] Baba, T., Yamaguchi, M. (2000) Stimulatory effect of regucalcin on proteolytic activity is impaired in the kidney cortex cytosol of rats with saline ingestion. *Mol. Cell. Biochem.* 206:1–6.

[78] Wang, K. K., Villalobo, A., Ronfogalis, B. D. (1989) Calmodulin-binding protein as calpain substrates. *Biochem. J.* 262:693–706.

[79] Croall, D. E., DeMartino, G. N. (1991) Calcium-activated neutral protease (calpain) system: structure, function, and regulation. *Physiol. Rev.* 71:813–847.

[80] Tamaoki, T., Nomoto, H., Takahashi, I., Kato, Y., Morimoto, M., Tomita, E. (1986) Staurosporine, a potent inhibitor of phospholipids/Ca^{2+}-dependent protein kinase. *Biochem. Biophys. Res. Commun.* 135: 397–402.

[81] Morooka, Y., Yamaguchi, M. (2002) Suppressive effect of endogenous regucalcin on deoxyribonucleic acid synthesis in the nuclei of rat renal cortex. *Mol. Cell. Biochem.* 229:157–162.

[82] Nakagawa, T., Sawada, N., Yamaguchi, M. (2005) Overexpression of regucalcin suppresses cell proliferation of cloned normal rat kidney proximal tubular epithelial NRK52E cells. *Int. J. Mol. Med.* 16:637–643.

[83] Charollais, R. H., Buquet, C., Mester, J. (1990) Butyrate blocks the accumulation of CDC2 mRNA in late G1 phase but inhibits both early and late G1 progression in chemically transformed mouse fibroblasts BP-A31. *J. Cell. Physiol.* 145:46–52.

[84] Meijer, L., Borgne, A., Mulner, O., Chong, J. P., Blow, J. J., Inagaki, N., et al. (1997) Biochemical and cellular effects of roscovitine, a potent and selective inhibitor of the cyclin-dependent kinases cdc2, cdk2 and cdk5. *Eur. J. Biochem.* 243:527–536.

[85] Singh, S. V., Herman-Antosiewice, A., Singh, A. V., Lew, K. L., Srivastava, S. K., Kamath, R., et al. (2004) Sulforaphane-induced G2/M phase cell cycle arrest involves Checkpoint kinase 2-mediated phosphorylation of cell division cycle 25C. *J. Biol. Chem.* 279:25813–25822.

[86] Park, Y. C., Lee, C. H., Kang, H. S., Chung, H. T., Kim, H. D. (1997) Wortmannin, a specific inhibitor of phosphatidylinositol-3-kinase, enhances LPS-induced NO production from murine peritoneal macrophages. *Biochem. Biophys. Res. Commun.* 240:692–696.

[87] Nakagawa, T., Yamaguchi, M. (2005) Overexpression of regucalcin suppresses apoptotic cell death in cloned normal rat kidney proximal tubular epithelial NRK52E cells: change in apoptosis-related gene expression. *J. Cell. Biochem.* 96:1274–1285.

[88] Liu, S., Shia, D., Liu, G., Chen, H., Liu, S., Hu, Y. (2000) Roles of Se and NO in apoptosis of hepatoma cells. *Life Sci.* 68:603–610.

[89] Cano-Abad, M. F., Villarroya, M., Garcia, A. G., Garbilan, N. H., Lopez, M. G. (2001) Calcium entry through L-type calcium channels causes mitochondrial disruption and chromaffin cell death. *J. Biol. Chem.* 276:39695–39704.

[90] Vogelstein, B., Lane, D., Levine, A. J. (2000) Surfing the p53 network. *Nature* 408:307–310.

[91] Widmann, C., Gibson, S., Johnson, G. L. (1988) Caspase-dependent cleavage of signaling proteins during apoptosis. A turn-off mechanism for anti-apoptotic signals. *J. Biol. Chem.* 273:7141–7147.

[92] Dieguez-Acuna, F. J., Polk, W. W., Ellis, M. E., Simmonds, P. L., Kushleika, J. V., Woods, J. S. (2004) Nuclear factor kappaB activity determines the sensitivity of kidney epithelial cells to apoptosis: implications for mercury-induced renal failure. *Toxicol. Sci.* 82:1114–1123.

[93] Nakagawa, T., Yamaguchi, M. (2005) Overexpression of regucalcin enhances its nuclear localization and suppresses L-type Ca^{2+} channel and calcium-sensing receptor mRNA expressions in cloned normal rat kidney proximal tubular epithelial NRK52E cells. *J. Cell. Biochem.* 99: 1064–1077.

[94] Ba, J., Friedman, P. A. (2004) Calcium-sensing receptor regulation of renal mineral ion transport. *Cell Calcium* 35:229–237.

[95] Wald, H., Ganty, H, Palmer, L. G., Popovtzer, M. M. (1998) Differential regulation of ROMK experession in kidney cortex and medulla by aldosterone and potassium. *Am. J. Physiol.* 275:F239–F245.

[96] Masilamani, S., Kim, G.-H., Mitchell, C., Wade, J. B., Knepper, M. A. (1999) Aldosterone-mediated regulation of EnaCα, β, γ subunit proteins in rat kidney. *J. Clin. Invest.* 104:R19–R23.

[97] Vanessa, S., Dabid, M., Frank, R., Marcelle, B., Pierre-Yves, M., Alain, V., et al. (2001) Short term effect of aldosterone on Na, K-ATPase cell surface expression in kidney collecting duct cells. *J. Biol. Chem.* 276: 47087 – 47093.

[98] Tanaka, T., Nangaku, M., Miyata, T., Inagi, R., Ohse, T., Ingelfinger, J. R., et al. (2004) Blockade of calcium influx through L-type calcium

channels attenuates mitochondrial injury and apoptosis in hypoxic renal tubular cells. *J. Am. Soc. Nephrol.* 15:2320–2333.

[99]   Ba, J., Friedman, P. A. (2004) Calcium-sensing receptor regulation of renal mineral ion transport. *Cell Calcium* 35:229–237.

[100] Gilbert, T., Leclerc, C., Moreau, M. (2011) Control of kidney development by calcium ions. *Biochimie* 93:2126–31.

[101] Fan, J. M., Ng, Y.-Y., Hill, P. A., Nikolic-Paterson, D. J., Mu, W., Atkins, R. C., et al. (1999) Transforming growth factor-β regulates tubular epithelial-myofibroblast transdifferentiation in vitro. *Kidney Int.* 56:1455–1467.

[102] Zhang, Y. Q., Kanzaki, M., Furukawa, M., Shibata, H., Ozeki, M., Kojima, I. (1999) Involvement of Smad proteins in the differentiation of pancreatic AR42J cells induced by activin A. *Diabetologia* 42:719–727.

[103] Hammar, E. B., Irminger, J.-G., Richenback, K., Parnaud, G., Ribaux, P., Bosco, D., et al. (2005) Activation of NF-κB by extracellular matrix is involved in spreading and glucose-stimulated insulin secretion of pancreatic beta cells. *J. Biol. Chem.* 280:30630–30637.

In: New Developments in Calcium …   ISBN: 978-1-62948-601-7
Editor: Masayoshi Yamaguchi   © 2014 Nova Science Publishers, Inc.

Chapter VII

Suppressive Role of Regucalcin in Cell Protein Synthesis: Involvement in Calcium Signaling

***Masayoshi Yamaguchi**[*]*
Department of Hematology and Medical Oncology,
Emory University School of Medicine,
Atlanta, GA, US

Abstract

Regucalcin is a novel calcium-binding protein that does not contain EF-hand motif of calcium-binding domain, which differs from calmodulin and other calcium-related proteins. The name, regucalcin, was proposed for this calcium-binding protein, which regulates various Ca^{2+}- or Ca^{2+}/calmodulin-dependent enzyme activations. Regucalcin plays a pivotal role as a regulatory protein in Ca^{2+}-signaling in many cell types, and it has been demonstrated to play a multifunctional role in cell regulation including the depression of apoptotic cell death and cell proliferation induced by various signaling factors in many cell types in

[*] Corresponding author: Masayoshi Yamaguchi, E-mail: yamamasa1155@yahoo.co.jp.

vitro. Cytoplasm regucalcin translocates to nucleus and suppresses nuclear deoxyribonucleic acid (DNA) and ribonucleic acid (RNA) synthesis. Moreover, regucalcin has been shown to have a suppressive effect on protein synthesis and a stimulatory effect on protein degradation. Regucalcin has been found to directly suppress aminoacyl-tRNA synthetase and activate Ca^{2+}-activated protease (calpains) in cells. Regucalcin may play a suppressive role in protein output in cells.

Keywords: Regucalcin, calcium signaling, protein synthesis, protein degradation, aminoacyl-tRNA synthetase, protease, calpains

Introduction

Regucalcin was discovered in 1978 as a calcium-binding protein that does not contain EF-hand motif as calcium-binding domain, which is found in many calcium-binding proteins [1-3]. The name, regucalcin, was proposed for this calcium-binding protein, which suppresses the activity of various enzymes that are activated by Ca^{2+} or Ca^{2+}/calmodulin [2, 3]. Regucalcin plays a pivotal role in the regulation of Ca^{2+} signaling as a suppressor protein [1-3]. The regucalcin gene is localized on X chromosome in consisting of seven exons and six introns [4, 5]. Regucalcin and its gene (*rgn*) are identified in over 15 species consisting of regucalcin family and the gene species are highly conserved in vertebrate species [6-8].

The regucalcin gene expression is regulated through various transcription factors (including AP-1, NF1-A1, RGPR-p117, β-catenin, SP1, and others), which are identified as the enhancer and suppressor, and this expression is regulated with hormonal stimulation and physiological state [8, 9-14]. Regucalcin has been demonstrated to play a multifunctional role in the regulation in various tissues and cell types [15-18]. Regucalcin plays a role in maintaining of intracellular Ca^{2+} levels due to activating various Ca^{2+} pump enzymes [15]. Cytoplasm regucalcin is localized into the nucleus and suppresses the nuclear Ca^{2+}- dependent protein kinase and protein phosphatase activities, Ca^{2+}-activated DNA fragmentation, and DNA and RNA synthesis [15, 16] in various type cells. Regucalcin has been proposed to play a pivotal role as a suppressor protein in various signal transductions to maintain cell homeostasis for stimuli [15, 16].

Moreover, regucalcin has been shown to have a suppressive effect on protein synthesis and a stimulatory effect on protein degradation. Regucalcin may play a suppressive role in protein output in cells. This chapter will introduce the findings concerning the molecular mechanism by which regucalcin suppresses protein synthesis and stimulates protein degradation in the involvement of calcium signaling.

Role of Regucalcin in Nuclear Regulation

Nuclear Localization of Regucalcin

Calmodulin stimulates DNA synthesis in liver cells [17] and the effect of calmodulin is mediated through α-adrenergic stimulation [18, 19]. Regucalcin has been shown to localize in the nuclei of rat liver [20, 21]. The nuclear regucalcin can bind on calmodulin-agarose beads [22]. Endogenous regucalcin has been shown to be present in the nuclei of rat liver using Western blotting analysis [23]. When isolated liver nuclei were incubated in the presence of exogenous regucalcin (50 µg/ml; 1.5 µM), potent band for regucalcin was found in the nucleus [23], supporting the view that the protein is translocated into the nucleus. A part of regucalcin, which is present in the cytoplasm of liver cells, is translocated to the nucleus [23]. Nuclear regucalcin translocation was not appreciably changed in the presence of ATP (2 mM), guanosine 5'-triphosphate (GTP, 2 mM), and calcium chloride (0.1 mM), suggesting that its translocation is not mediated through nuclear localization signal [23]. ATP and GTP are required for nuclear import of proteins that are localized in the nuclei. ATP or GTP did not regulate the translocation of regucalcin into liver nuclei. The nuclear localization of regucalcin was independently of Ca^{2+}.

Nuclear protein transport is blocked in the presence of the lectin wheat germ agglutinin (WGA) [24]. Nuclear regucalcin translocation was not appreciably changed in the presence of WGA in the reaction mixture [23]. This finding suggests that the nuclear translocation of regucalcin is not related to nuclear localization signal that is responsible for selection for intranuclear active transport. The molecular weight of regucalcin is about 33 kDa [6]. Regucalcin may be passively transported to the nucleus through nuclear pore in liver cells.

Regucalcin has also been shown to localize in the nuclei of the cloned normal rat kidney proximal tubular epithelial NRK52E cells with immunocytochemical analysis [25]. The nuclear localization of regucalcin is enhanced through hormonal signaling process that is involved in protein kinase C [25].

Regucalcin has been shown to bind proteins in isolated rat liver nuclei using Far-Western blot analysis [26]. The results of the Far-Western analysis showed the existence of protein components that bind to regucalcin in the nucleus isolated from rat liver [26]. Regucalcin has also been demonstrated to bind DNA using Western blot analysis for regucalcin with DNA cellulose-binding assay [26].

Regucalcin Suppresses Nuclear Enzyme Activity

Isolated rat liver nucleus contains a DNA endonuclease activity dependent upon Ca^{2+} and Ca^{2+} results in extensive DNA hydrolysis [27]. The Ca^{2+} dependence of this endogenous DNA fragmentation process is based on DNA endonuclease activity dependent upon sub-micromolar Ca^{2+} when the nucleus is reconstituted with NAD^{+} and ATP [27]. This endogenous endonuclease activity may be responsible for the DNA fragmentation occurring during programmed cell death (apoptosis) and certain forms of chemically induced cell killing [27, 28]. Regucalcin has shown to have a suppressive effect on Ca^{2+}-activated DNA fragmentation in isolated rat liver nuclei [29]. The presence of regucalcin (0.5-2.0 µM) completely depressed the activation of liver nuclear DNA fragmentation when 10 µM Ca^{2+} was added [29]. Regucalcin can bind to nuclear proteins in the absence or presence of 1.0 mM Ca^{2+} [30]. Presumably, regucalcin has a direct suppressive effect on DNA endonuclease in liver nuclei. Several studies have shown that Ca^{2+} plays an important role in the regulation of nuclear functions [31, 32]. A sustained increase in cytosolic Ca^{2+} level precedes the activation of DNA fragmentation that is characteristic of programmed cell death (apoptosis) and in certain forms of chemically induced cell killing [27, 28]. Regucalcin may suppress apoptosis, which is induced through the activation of DNA fragmentation.

Smal GTPase Ran (ras-related nuclear protein) is required for protein export from the nucleus and protein import into the nucleus [33]. The role of regucalcin in the regulation of GTPase activity in the nuclei of rat liver has been shown [34]. GTPase is found in the nuclei isolated from rat liver [34]. Liver nuclear GTPase activity was increased after calcium addition in the enzyme reaction mixture, and this increase was not seen in the presence of trifluoperazine (TFP), an antagonist of calmodulin [34]. The presence of exogenous regucalcin (0.5 µM) in the enzyme reaction mixture caused an inhibitory effect on GTPase activity in liver nucleus [34]. This effect was also seen in the presence of ethyleneglycol bis (2-amino-ethylether)-*N, N, N', N'*-tetraacetic acid (EGTA), a chelator of Ca^{2+}. The inhibitory effect of regucalcin on liver nuclear GTPase activity was independent of Ca^{2+}/calmodulin in the nucleus. Regucalcin has been shown to have an inhibitory effect on the activation of enzymes by Ca^{2+}/calmodulin due to binding Ca^{2+} and/or calmodulin [35-38]. Regucalcin may directly inhibit GTPase activity in liver nucleus.

Regucalcin may be able to regulate a process of signal transduction from the cytoplasm to nucleus in liver cells.

This process is mediated through various protein kinases. The role of regucalcin in the regulation of Ca^{2+}-dependent protein kinase and protein tyrosine phosphatase activities in isolated liver nucleus has been examined [21, 23, 39]. Nuclear Ca^{2+}-dependent protein kinase and protein tyrosine phosphatase activities were increased in the presence of anti-regucalcin monoclonal antibody in the enzyme reaction mixture, and these increases were completely depressed after addition of regucalcin [23]. Endogenous regucalcin has been demonstrated to have a suppressive effect on the enhancement of protein kinase activity with a proliferation of liver cells [40]. Protein kinase activity is enhanced in the cytosol and nucleus of regenerating rat liver with proliferating cells *in vivo* [40]. Regucalcin has been shown to have a suppressive effect on tyrosine kinase, protein kinase C, and Ca^{2+}/calmodulin-dependent protein kinase in the cytoplasm and nucleus of regenerating rat liver [35, 36, 40].

Phosphatase activity toward phosphotyrosine, phosphoserine, and phosphothreonine is found in the liver nucleus [21]. The addition of regucalcin in the enzyme reaction mixture caused a significant decrease in phosphatase activity toward phosphotyrosine, phosphoserine, and phosphothreonine in the liver nuclei [21]. Liver nuclear phosphatase activity toward phosphoamino acids was assayed using 5-6 mg of nuclear protein per milliliter of reaction mixture; it contained 275-330 ng of nuclear regucalcin [21]. The concentration of endogenous nuclear regucalcin was estimated about 82.4-98.8 nM. Further addition of exogenous regucalcin (0.25 μM) caused a significant decrease in nuclear phosphatase activity, although the effect was saturated with increasing concentrations of regucalcin (0.5 μM) [21]. Nuclear protein phosphatase activity was elevated in the presence of anti-regucalcin monoclonal antibody (25 and 50 ng/ml of reaction mixture) [21]. This elevation was completely depressed after addition of regucalcin. Endogenous regucalcin may regulate activity of various protein phosphatases in liver nucleus.

Regucalcin Suppresses Nuclear DNA and RNA Synthesis

Regucalcin has been shown to have an inhibitory effect on DNA synthesis activity in the nuclei of normal rat liver [41]. The inhibitory effect of regucalcin was seen in the presence of EGTA, a chelator of Ca^{2+}, in the reaction mixture. Ca^{2+} is present in liver nucleus [41].

Liver nuclear DNA synthesis activity was increased in the presence of EGTA in the reaction mixture, suggesting that Ca^{2+} suppresses DNA synthesis activity in the nucleus. The effect of regucalcin in suppressing nuclear DNA synthesis may be not related to Ca^{2+} in liver nucleus.

Liver nuclear DNA synthesis has been shown to stimulate in regenerating rat liver *in vivo* [41]. Nuclear DNA synthesis was markedly increased at 1 day after hepatectomy, and this increase was also seen at 3 days [41]. Nuclear DNA synthesis was enhanced in the presence of EGTA, Ca^{2+} chelator in the incubation mixture [41]. The presence of Ca^{2+} (1.0 − 25 µM) caused a significant decrease in the nuclear DNA synthesis of normal rat liver. Regucalcin (0.25 and 0.5 µM) caused an inhibition of nuclear DNA synthesis of normal rat liver [41]. This inhibition was also seen in the presence of Ca^{2+}. The inhibitory effect of regucalcin was remarkable in regenerating rat liver nuclei in comparison with that of normal rat liver [41]. Thus, regucalcin has been shown to have a suppressive effect on nuclear DNA synthesis in regenerating rat liver with proliferating cells.

Regucalcin has been also shown to have a suppressive effect on DNA synthesis activity in the nuclei isolated from rat renal cortex [43]. The addition of regucalcin (0.1 − 0.5 µM) in the reaction mixture had an inhibitory effect on nuclear DNA synthesis activity independent on calcium addition [44]. The presence of anti-regucalcin monoclonal antibody (10-50 ng/ml) in the reaction mixture caused a significant increase in nuclear DNA synthesis activity [44]. This increase was completely depressed in the presence of regucalcin (0.5 µM). Endogenous regucalcin has been found to have a suppressive effect on DNA synthesis in the nuclei of rat renal cortex [44].

Regucalcin has been shown to have a suppressive effect on RNA synthesis in the nuclei isolated from normal rat liver *in vitro* [44] and regenerating rat liver *in vivo* [45]. Ca^{2+} has a stimulatory effect on RNA synthesis in liver nucleus [46, 47]. This effect may be partly mediated through Ca^{2+}-dependent protein kinase [46, 47]. The stimulatory effect of Ca^{2+} on nuclear RNA synthesis activity was completely depressed in the presence of regucalcin [44, 45]. The addition of Ca^{2+} with higher concentrations has been reported to have an inhibitory effect of nuclear RNA synthesis in rat liver cells [46].

Inactivation of RNA polymerase III transcription has been shown to be calcium dependent [47]. The effect of regucalcin in decreasing nuclear RNA synthesis activity in normal rat liver was not seen in the presence of α-amanitin, an inhibitor of RNA polymerase II and III [45], suggesting that the suppressive effect of regucalcin is partly resulted from its inhibitory action on RNA polymerase II and III [44, 45]. Regucalcin has been shown to inhibit Ca^{2+}-dependent protein kinases in rat liver nucleus [40]. Presumably, the effect of regucalcin in decreasing RNA synthesis activity in liver nucleus is partly involved in its inhibitory action on the activities of both RNA polymerase II and III and Ca^{2+}-dependent protein kinases.

As described above, regucalcin has been found to have a suppressive effect on liver nuclear DNA and RNA synthesis [45-46]. The mechanism by which regucalcin inhibits nuclear DNA and RNA synthesis has been poorly understood. Regucalcin may have an inhibitory effect on DNA and RNA polymerase activity. However, the effect of regucalcin in inhibiting nuclear RNA synthesis activity is observed in the presence of α-amanitin, an inhibitor of RNA polymerase II. Regucalcin can directly bind DNA [26], and it may reveal a suppressive effect on nuclear DNA and RNA synthesis. The suppressive effect of regucalcin on liver nuclear DNA and RNA synthesis may partly influence on protein output in liver cells.

Regucalcin Suppresses Protein Synthesis

Regucalcin has been shown to have a regulatory effect on protein synthesis and protein degradation, suggesting that regucalcin plays a role in the regulation of protein turnover in cells. Protein synthesis is depressed in a variety of eukaryotic cell types exposed to conditions depleting Ca^{2+} but not Mg^{2+} [48]. It has also been proposed that hormones (vasopressin and α-adrenergic agonist), which have been known to mobilize sequestered Ca^{2+} within liver cells, inhibit amino acid incorporation by influencing a Ca^{2+} requirement associated with protein synthesis [49]. Moreover, vasopressin inhibits the rate of protein synthesis in isolated hepatocytes partially depleted of Ca^{2+} [50]. These investigations propose the hypothesis that a sequestered pool of intracellular Ca^{2+} is required for the maintenance of high rates of protein synthesis in liver cells. On the other hand, vasopressin and α-adrenergic agonist cause an increase in the intracellular free Ca^{2+} concentration of hepatocytes that are not depleted of Ca^{2+} [51, 52]. However, the effects of intracellular Ca^{2+} on hepatic protein synthesis have been well undefined. It may be important to clarify the effect of Ca^{2+} on hepatic protein synthesis *in vitro*.

It has been demonstrated that Ca^{2+}, of various metals, can uniquely inhibit *in vitro* protein synthesis using the 5500 g supernatant fraction (the microsomes and cytosl) of rat liver homogenate [53]. This inhibition was seen after addition of 1.0 μM Ca^{2+}. Ca^{2+} addition caused a remarkable decrease of the activity of aminoacyl (leucyl)-tRNA synthetase, which is a rate-limiting enzyme of protein synthesis at translational process, in hepatic cytosol [53].

Ca^{2+} may directly inhibit hepatic protein synthesis in subcellular fraction of liver cells. Ca^{2+} is required for protein synthesis in hepatocytes exposed to conditions depleting the cation [53]. Calmodulin, which can amplify Ca^{2+} effects on enzymes, did not have an appreciable effect on *in vitro* protein synthesis using the 5500 g supernatant fraction of liver homogenate in the presence of Ca^{2+} (10 μM) [53]. The activity of aminoacyl-tRNA synthetase in hepatic cytosol was not altered through calmodulin [54]. The mechanism by which Ca^{2+} is required in protein synthesis of hepatocytes depleted of Ca^{2+} may be complex.

The role of regucalcin in the regulation of *in vitro* protein synthesis using the 5500 g supernatant fraction of rat liver homogenate has been investigated [53]. Regucalcin caused a remarkable inhibition of hepatic protein synthesis *in vitro* [53]. Regucalcin could not reverse the Ca^{2+}-induced inhibition of protein synthesis [53], although regucalcin has been shown to reverse the effect of Ca^{2+} on many enzymes in liver cells. Since regucalcin can bind to liver cytosolic proteins and its binding is slightly enhanced with the coexistence of 0.1 μM Ca^{2+} [50], Ca^{2+}-binding regucalcin and/or Ca^{2+} free regucalcin may be able to inhibit hepatic protein synthesis. In fact, the presence of regucalcin (1 and 2 μM) could fairly decrease hepatic protein synthesis that was reduced after addition of 10 μM Ca^{2+} [53]. Regucalcin may play a role in the regulation of protein synthesis independently of Ca^{2+} in liver cells.

Regucalcin has been shown to inhibit hepatic aminoacyl-tRNA synthase activity [53]. The inhibitory effect of regucalcin was seen in the presence of Ca^{2+} (10 μM). The inhibitory effect of regucalcin on hepatic protein synthesis may be partly based on a remarkable decrease of aminoacyl-tRNA synthetase activity. Regucalcin may bind aminoacyl (leucyl)-tRNA synthetase in hepatic cytosol, since iodinated regucalcin can bind the proteins in hepatic cytosol [30].

The role of endogenous regucalcin on protein synthesis has been examined using anti-regucalcin monoclonal antibody, moreover [54]. The presence of anti-regucalcin monoclonal antibody in the reaction mixture caused a significant increase in protein synthesis and [^{3}H] leucyl-tRNA synthetase activity in normal rat liver. These increases were completely depressed after addition of exogenous regucalcin (1.0 µM). Liver cytosol contained about 16 µg of regucalcin per 1 mg of the cytosolic protein; the reaction mixture contained about 0.17-0.19 µM of endogenous regucalcin, because the cytosolic protein in the range 360 − 390 µg were added into the mixture of 1.0 ml. Endogenous regucalcinin may have a suppressive effect on protein synthesis in liver cells.

Hepatic protein synthesis has been shown to enhance in regenerating rat liver, which induces a proliferation of liver cells after partial hepatectomy *in vivo* [54]. This enhancement was reamarkable at 24 and 48 hours after partial hepatectomy. Hepatic protein synthesis in regenerating liver was further enhanced in the presence of anti-regucalcin monoclonal antibody in the reaction mixture. Endogenous regucalcin may play a suppressive role on the enhancement of protein synthesis in regenerating liver.

Regucalcin Stimulates Protein Degradation

Evidence for the role of Ca^{2+}-activated protease (calpains) is implicated in signal transduction [55]. Two neutral Ca^{2+}-requiring proteinases, differing in molecular size, have been isolated from rabbit liver cytosol [56]. Both are recovered as inactive proenzymes that can be converted to the active forms by high (0.1-1.0 mM) concentrations of Ca^{2+} in the absence of substrate or, in the presence of a protein substrate, by low (1-5 µM) concentrations of Ca^{2+} [56]. The activated proteinases required only 1-5 µM Ca^{2+} for maximal activity [57].

The addition of regucalcin (0.25 − 2.0 µM) into the enzyme reaction mixture has been found to induce a remarkable increase of neutral proteinase activity in the presence of 5.0 µM Ca^{2+} [57]. The effect of regucalcin was seen at 0.25 µM. Regucalcin may activate Ca^{2+}-requiring proteinase in rat liver cytosol. The effect of regucalcin in increasing liver cytosolic proteinase activity was also seen in the absence of Ca^{2+} [57]. This increase was remarkable as compared with that of Ca^{2+} addition. Regucalcin may activate Ca^{2+}-not requiring neutral proteinase in rat liver cytosol. Regucalcin has a reversible effect on the activation of various enzymes by Ca^{2+} and/or calmodulin [35-37]. The finding, that regucalcin can activate neutral proteinase in rat liver cytosol in the presence or absence of Ca^{2+}, was novel. The activating effect of regucalcin on liver cytosolic proteinase activity was not seen in the presence of anti-regucalcin antiserum in the enzyme reaction mixture [57], suggesting a role of endogenous regucalcin in the activation of cytosolic proteinase.

The activating effect of regucalcin on neutral proteases in the liver cytosol has been characterized [58]. Leupeptin is a potent inhibitor of SH-proteinase. The regucalcin-increased protenase activity was inhibited in the presence of leupeptin in the enzyme reaction mixture [58]. Regucalcin may activate neutral cysteinyl-proteinase in the liver cytosol.

The effect of regucalcin in increasing proteolytic activity in rat liver cytosol was not depressed in the presence of diisopropylfluorophosphate (DFP), an inhibitor of serine-protease, although DFP alone had an inhibitory effect on the proteolytic activity [58]. Regucalcin did not act on serine-proteases in liver cytosol. The effect of regucalcin was also seen in the presence of a chelator of metal ions, suggesting that regucalcin does not activate metal-related proteases in liver cytosol.

The proteolytic activity in liver cytosol was markedly increased after addition of dithiothreitol (DTT), a protecting reagent for thiol (SH) group, in the enzyme reaction mixture, and this increase was completely inhibited in the presence of N-ethylmaleiimide (NEM), a SH group-modifying agent [58]. The effect of regucalcin in increasing proteolytic acitivity was seen in the presence of DTT in liver cytosol, although it was completely inhibited in the presence of NEM [58]. Regucalcin may act on the SH group of cysteinyl-proteases in liver cytosol.

Two forms of Ca^{2+}-activated neutral proteases (calpains) have been identified in hepatocytes and other cells, *m*-calpaine and μ-calpaine [55]. Isolated μ-calpain requires micromolar concentrations of Ca^{2+} for activation, while *m*-calpain requires millimolar concentrations [59]. The activation of proteases in rabbit liver cytosol is required only 1-5 μM Ca^{2+} for maximal activity [58]. Ca^{2+} (10 μM)-increased proteolytic activity in rat liver cytosol was inhibited in the presence of NEM, indicating that neutral proteases (including calpains), which are activated by Ca^{2+}, exist in liver cytoplasm [58].

The activity of calpaines, which are thiol protease [60], increases in hepatocytes following addition of ATP [61]. The proteolytic activity in liver cytosol was decreased after the addition of calpastatin, a specific inhibitory of calpains, indicating the existence of calpains in the cytosol [58]. Regucalcin-increased proteolytic activity was depressed in the presence of calpastatin.

Regucalcin may be an activator for calpains. *m*-Calpain isolated from rabbit skeletal muscle was activated by regucalcin independent on Ca^{2+}. This activation was inhibited after addition of NEM [58]. Regucalcin acts on the SH group in *m*-calpain. Presumably, regucalcin may be able to activate both *m*- and μ-calpains.

The role of regucalcin in the regulation of neutral proteolytic activity in rat kidney cortex cytosol has also been examined [62, 63]. Regucalcin had an activating effect on neutral proteolytic activity in the kidney cortex cytosol [62]. This increase was depressed in the presence of anti-regucalcin monoclonal antibody, supporting the view that endogenous regucalcin plays a role as activator of proteases in the renal cortex ctrtoscl [62].

The activating effect of regucalcin on proteases is seen in the concentrations of 0.01 − 0.25 μM [62]. The concentration of regucalcin in rat kidney tissue has been found to be present about 5.3 μM. Regucalcin plays a physiological role in the activation of thiol proteases in renal cortex cells.

The effect of regucalcin on proteolytic activity was not altered in the presence of calcium chloride (0.01 and 1.0 mM) or EGTA (1.0 mM), indicating that the effect of regucalcin was independent on Ca^{2+}. Regucalcin may activate both calpaines and other proteases in the kidney cortex cytosol. Regucalcin has also been shown to activate SH proteases in rat renal cortex cytosol. Presumably, regucalcin acts on the SH-groups of protease in rat renal cortex cytosol.

Regucalcin uniquely activates thiol proteases independent on Ca^{2+} in the liver cytosol, whereas it has no effect on serine proteases and metalloproteases [57, 58]. Such an effect of regucalcin was also found in the renal cortex cytosol [63]. Regucalcin may be an activator on thiol proteases in liver and kidney cells. Regucalcin also plays a role as an activator in other tissues that express regucalcin.

Calpains are ubiquitous, non-lysosomal, calcium-dependent proteases that may play important roles in Ca^{2+}-mediated intracellular processes [55, 56, 61]. The ability of calpain to alter the limited proteolysis, the activity or function of numerous cytoskeletal proteins, protein kinases, receptors and transcription factors suggests an involvement of protease in Ca^{2-}-regulated cellular functions [60].

Calpains may also play an integral role in modulating the activity of protein kinase C, a key protein in many signal transduction processes [64], since they convert the native Ca^{2+}/phospholipids-dependent kinase to a soluble form that does not require Ca^{2+} or phospholipids for activity [65].

Regucalcin can increase the activity of thiol proteases including calpain in rat liver and renal cortex cytosols. Regucalcin may play a pivotal role in the regulation of cellular functions related to Ca^{2+} mediated through thiol proteases. Presumably, regucalcin plays an important role in the regulation of signal transduction that is involved in proteases.

As summarized in Figure 1, regucalcin may play a role as suppressor in the enhancement of cytoplasm protein levels due to inhibiting protein synthesis and activating protein degradation in normal and proliferating cells. Regucalcin may regulate protein turnover in cells. Further studies are expected.

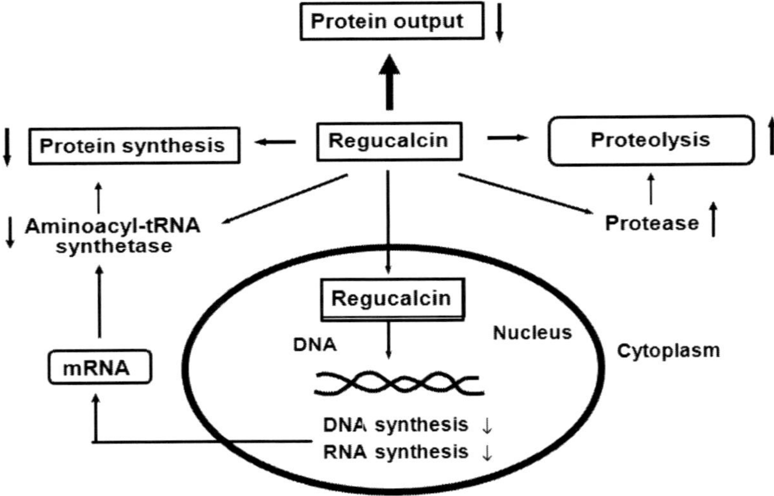

Figure 1. Regucalcin has a suppressive effect on protein output due to depressing protein synthesis and stimulating protein degradation in normal and proliferating liver cells. This effect may be based on suppressing of nuclear DNA and RNA synthesis and aminoacyl t-RNA synthetase, a rate- limited enzyme of translational process of protein synthesis, and stimulating of proteolysis that is mediated through activation of thiol protease. Regucalcin may regulate protein turnover in the cells. Regucalcin may play a pathophysiological role in suppression of the overexpression of protein output in proliferating cells.

Conclusion

Regucalcin plays a pivotal role in cell regulation as a suppressor protein in signal transduction for various stimuli in various type cells [15, 16].

Regucalcin has been shown to have a suppressive effect on cell proliferation and apoptosis that are mediated through various hormones, cytokine and other factors. Moreover, regucalcin has been demonstrated to suppress protein synthesis and stimulate proteolysis in cells. Suppressive effect of regucalcin on protein synthesis is mediated through inhibiting of transcriptional and translational process. Stimulatory effect of regucalcin on proteolysis is mediated through activation of thiol protease. Suppressive effect of regucalcin on protein output in cells is revealed in normal and proliferating liver cells. Regucalcin may play a role as a regulatory factor in protein turnover in cells. Especially, regucalcin may have a suppressive effect on overexpression of protein output in proliferating cells to maintain cell homeostasis.

This effect of regucalcin may play a role in depression of the development of carcinogenesis that enhances cell protein output, if regucalcin expression is down-regulated in the development of carcinogenesis [64]. The targeting of regucalcin gene may have usefulness in suppression of carcinogenesis.

References

[1]   Yamaguchi, M., Yamamoto, T. (1978) Purification of calcium binding substance from soluble fraction of normal rat liver. *Chem. Pharm. Bull.* 26: 1915-1918.

[2]   Yamaguchi, M., Mori, S. (1988) Effect of Ca^{2+} and Zn^{2+} on 5'-nucleotidase activity in rat liver plasma membranes: Hepatic calcium-binding protein (regucalcin) reverses the Ca^{2+} effect. *Chem. Pharm. Bull.* 36: 321-325.

[3]   Yamaguchi, M. (1992) A novel Ca^{2+}-binding protein regucalcin and calcium inhibition. Regulatory role in liver cell function. In: *Calcium Inhibition.* K. Kohama (ed.) Japan. Sci. Soc. Press, Tokyo and CRC Press, Boca Raton pp. 19-41.

[4]   Shimokawa, N., Matsuda, Y., Yamaguchi, M. (1995) Genomic cloning and chromosomal assignment of rat regucalcin gene. *Mol. Cell. Biochem.* 151:157-163.

[5]   Thiselton, D. L., McDowall, J., Brandau, O., Ramser, J., d'Esposito, F., Bhattacharga, S. S., Ross, M. T., Hardcastle, A. J., Meindl, A. (2002) An integrated, functionally annotated gene map of the DXS8026-ELK1 internal on human Xp11.3-Xp11.23: Potential hotspot for neurogenetic disorders. *Genomics* 79: 560-572.

[6]   Shimokawa, N., Yamaguchi, M. (1993) Molecular cloning and sequencing of the cDNA coding for a calcium-binding protein regucalcin from rat liver. *FEBS Lett.* 327: 251-255.

[7]   Misawa, H., Yamaguchi, M. (2000) The gene of Ca^{2+}-binding protein regucalcin is highly conserved in vertebrate species. *Int. J. Mol. Med.* 6: 191-196.

[8]   Yamaguchi, M. (2011) The transcriptional regulation of regucalcin gene expression. *Mol. Cell. Biochem.* 346:147-171.

[9]   Yamaguchi, M., Makino, R., Shimokawa, N. (1996) The 5'end seguences and exon organization in rat regucalcin gene. *Mol. Cell. Biochem.* 165: 145-150.

[10]   Murata, T., Yamaguchi, M. (1998) Ca^{2+} administration stimulates the binding of AP-1 factor to the 5'-flanking region of the rat gene for the Ca^{2+}-binding protein regucalcin. *Biochem. J.* 329: 157-163.

[11]   Murata, T., Yamaguchi, M. (1999) Promoter characterization of the rat gene for Ca^{2+}-binding protein regucalcin. Transcriptional regulation by signaling factors. *J. Biol. Chem.* 274: 1277-1285.

[12]   Misawa, H., Yamaguchi, M. (2001) Molecular cloning and sequencing of the cDNA coding for a novel regucalcin gene promoter region-related protein in rat, mouse and human liver. *Int. J. Mol. Med.* 8: 513-520.

[13]   Yamaguchi, M. (2009) Novel protein RGPR-p117: its role as the regucalcin gene transcription factor. *Mol. Cell. Biochem.* 327: 53-63.

[14]   Nejak-Bowen, K. N., Zeng, G., Tan, X., Cieply, B., Monga, S. P. (2009) β-Catenin regulates vitamic C biosynthesis and cell survival in murine liver. *J. Biol. Chem.* 284:28115-28127.

[15]   Yamaguchi, M. (2005) Role of regucalcin in maintaining cell homeostasis and function (Review). *Int. J. Mol. Med.,* 15: 371-389.

[16]   Yamaguchi, M. (2011) Regucalcin and cell regulation: role as a supressor in cell signaling. *Mol. Cell. Biochem.,* 353:101-137.

[17]   Boynton, A. L., Whitfield, J. F., MacManus, J. P. (1980) Calmodulin stimulates DNA synthesis by rat liver cells. *Biochem. Biohys. Res. Commun.* 95:745-749.

[18]   Cruise, J., Houck, K. A., Michalopoulos, G. K. (1985) Induction of DNA synthesis in cultured rat hepatocytes through stimulation of alpha 1 adrenoreceptor by norepinephrine. *Science* 227:749-751.

[19]   Pujol, M. J., Soriano, M., Alique, R., Carafoli, E., Bachs, O. (1989) Effect of alpha-adrenergic blockers on calmodulin associate with the nuclear matrix of rat liver cells during proliferative activation. *J. Biol. Chem.* 264:18863-18865.

[20] Tsurusaki, Y., Misawa, H., Yamaguchi, M. (2000) Translocation of regucalcin to rat liver nucleus: Involvement of nuclear protein kinase and protein phosphatase regulation. *Int. J. Mol. Med.* 6: 655-660.

[21] Omura, M. and Yamaguchi, M. (1999) Regulation of protein phosphatase activity by regucalcin localization in rat liver nuclei. *J. Cell. Biochem.* 75: 437-445.

[22] Omura, M., Yamaguchi, M. (1998) Inhibition of Ca^{2+}/ calmodulin-dependent phosphatase activity by regucalcin in rat liver cytosol: Involvement of calmoduling binding. *J. Cell. Biochem.* 71: 140-148.

[23] Tsurusaki, Y., Misawa, H., Yamaguchi, M. (2000) Translocation of regucalcin to rat liver nucleus: Involvement of nuclear protein kinase and protein phosphatase regulation. *Int. J. Mol. Med.* 6: 655-660.

[24] Csermely, P., Schnaider, T., Szanto, I. (1995) Signalling and transport through the nuclear membrane. *Biochim. Biophys. Acta* 1241: 425-452.

[25] Nakagawa, T., Yamaguchi, M. (2008) Nuclear localization of regucalcin is enhanced in culture with protein kinase C activation in cloned normal rat kidney proximal tubular epithelial NRK52E cells. *Int. J. Mol. Med.* 21:605-610.

[26] Tsurusaki, Y., Yamaguchi, M. (2004) Role of regucalcin in liver nuclear function: Binding of regucalcin to nuclear protein or DNA and modulation of tumor-related gene expression. *Int. J. Mol. Med.* 14:277-281.

[27] Jones, D. P., McConkey, D. J., Nicotera, P., and Orrenius, S. (1989) Calcium- activated DNA fragmentation in rat liver nuclei. *J. Biol. Chem.* 264: 6398-6403.

[28] Farber, J. L. (1981) The role of calcium in cell death. *Life Sci.* 29: 1289-1295.

[29] Yamaguchi, M., Sakurai, T. (1991) Inhibitory effect of calcium-binding protein regucalcin on Ca^{2+}-activated DNA fragmentation in rat liver nuclei. *FEBS Lett.*, 279:281-284.

[30] Yamaguchi, M., Mori, S., Kato, S. (1988) Calcium-binding protein regucalcin is an activator of $(Ca^{2+}-Mg^{2+})$-adenosine triphosphatase in the plasma membranes of rat liver. *Chem. Pharm. Bull.* 36: 3532-3539.

[31] Nicotera, P., McConkey, D. J., Jones, D. P., Orrenius, S. (1989) ATP stimulates Ca^{2+} uptake and increases the free Ca^{2+} concentration in isolated rat liver nuclei. *Proc. Natl. Acad. Sci. US* 86: 453-457.

[32] Bachs, O., Carafolli, E. (1995) Calmodulin and calmodulin-binding proteins in liver cell nuclei. *J. Biol. Chem.* 262, 10786-10790.

[33] Moroianu, J., Blobel, G. (1995) Protein export from the nucleus requires the GTPase Ran and GTP hydrolysis. *Proc. Natl. Acad. Sci. US* 92:4318-4322.

[34] Tsurusaki, Y., Yamaguchi, M. (2001) Suppressive effect of endogenous regucalcin guanosine triphosphatase activity in rat liver nucleus. *Biol. Pharm. Bull.*, 24:958-961.

[35] Mori, S., Yamaguchi, M. (1990) Hepatic calcium-binding protein regucalcin decreases Ca^{2+}/calmodulin-dependent protein kinase activity in rat liver cytosol. *Chem. Pharm. Bull.* 38: 2216-2218.

[36] Kurota, H., Yamaguchi, M. (1997) Inhibitory effect of regucalcin on Ca^{2+}/calmodulin-dependent protein kinase activity in rat renal cortex cytosol. *Mol. Cell. Biochem.* 177: 239-243.

[37] Hamano, T., Hanahisa, Y., Yamaguchi, M. (1999) Inhibitory effect of regucalcin on Ca^{2+}-dependent protein kinase activity in rat brain cytosol: Involvement of endogenous regucalcin. *Brain Res. Brain* 50: 187-192.

[38] Hamano, T., Yamaguchi, M. (2001) Inhibitory role of regucalcin in the regulation of Ca^{2+}-dependent protein kinase activity in rat brain tissues. *J. Neurol. Sci.* 183: 33-38.

[39] Omura, M., Yamaguchi, M. (1999) Enhancement of neutral phosphatase activity in the cytosol and nuclei of regenerating rat liver: Role of endogenous regucalcin. *J. Cell. Biochem.* 73: 332-341.

[40] Katsumata, T., Yamaguchi, M. (1998) Inhibitory effect of calcium-binding protein regucalcin on protein kinase activity in the uclei of regenerating rat liver. *J. Cell. Biochem.* 71:569-576.

[41] Yamaguchi, M., Kanayama, Y. (1996) Calcium-binding protein regucalcin inhibits deoxyribonucleic acid synthesis in the nuclei of regenerating rat liver. *Mol. Cell. Biochem.* 162:121-126.

[42] Tsurusaki, Y., Yamaguchi, M. (2002) Suppressive role of endogenous regucalcin in the enhancement of deoxyribonucleic acid synthesis activity in the nucleus of regenerating rat liver. *J. Cell. Biochem.* 85: 516-5.

[43] Morooka, Y., Yamaguchi, M. (2002) Suppressive effect of endogenous regucalcin on deoxyribonucleic acid synthesis in the nuclei of rat renal cortex. *Mol. Cell. Biochem.* 229: 157-162.

[44] Yamaguchi, M., Ueoka, S. (1997) Inhibitory effect of calcium-binding protein regucalcin on ribonucleic acid synthesis in isolated rat liver nuclei. *Mol. Cell. Biochem.* 173: 169-175.

[45] Tsurusaki, Y., Yamaguchi, M. (2002) Role of endogenous regucalcin in nuclear regulation of regenerating rat liver: Suppression of the enhanced ribonucleic acid synthesis activity. *J. Cell. Biochem.* 87: 450-457.

[46] Pardo, J. P., Fernandez, F. (1982) Effect of calcium and calmodulin on RNA synthesis in isolated nuclei from rat liver cells. *FEBS Lett.* 143: 157-160.

[47] Sturges, M. R., Peck, L. J. (1994) Calcium-dependent inactivation of RNA polymerase III transcription. *J. Biol. Chem.* 269:5712-5719.

[48] Brostrom, C. O., Bocckino, S. B., Brostrom, M. A., Galuska, E. M. (1983) Identification of a Ca^{2+} requirement for protein synthesis in eukaryotic cells. *J. Biol. Chem.* 258:14390-14399.

[49] Brostrom, C. O., Bocckino, S. B., Brostrom, M. A., Galuska, E. M. (1986) Regulation of protein synthesis in isolated hepatocytes by calcium-mobilizing hormones. *Mol. Pharmacol.* 29:104-111.

[50] Menaya, J., Parvilla, R., Ayuso, M. S. (1988) Effect of vasopressin on the regulation of protein synthesis initiation in liver cells. *Biochem.* 254: 773-779.

[51] Thomas, A. P., Alexander, J., Williamson, J. R. (1984) Relationship between inositol polyphosphate production and the increase of cytosolic free Ca^{2+} induced by vasopressin in isolated hepatocytes. *J. Biol. Chem.* 259: 5574-5584.

[52] Berthon, B., Binet, A., Mauger, J.-P., Claret, M. (1984) Cytosolic free Ca^{2+} in isolated rat hepatocytes as measured by quin 2. Effects of noradrenalin and vasopressin. *FEBS Lett.* 167:19-24.

[53] Yamaguchi, M., Mori, S. (1990) Effect of calcium-binding protein regucalcin on hepatic protein synthesis: Inhibition of aminoacyl-tRNA synthetase activity. *Mol. Cell. Biochem.* 99:25-32.

[54] Tsurusaki, Y., Yamaguchi, M. (2000) Suppressive effect of endogenous regucalcin on the enhancement of protein synthesis and aminoacyl-tRNA synthetase activity in regenerating rat liver. *Int. J. Mol. Med.*, 6: 295-299.

[55] Wang, K. K., Villalobo, A., Ronfogalis, B. D. (1989) Calmodulin-binding protein as calpain substrates. *Biochem. J.* 262:693-706.

[56] Pontremoli, S., Melloni, E., Salamino, F., Sparator, B., Michetti, M., Horecker, B. L. (1984) Cytosolic Ca^{2+}-dependent neutral proteinases from rabbit liver: Activation of the proenzymes by Ca^{2+} and substrate. *Proc. Natl. Acad. Sci. US* 81:53-56.

[57] Yamaguchi, M., Tai, H. (1992) Calcium-binding protein regucalcin increases calcium-independent proteolytic activity in rat liver cytosol. *Mol. Cell. Biochem.*, 12:89-95.

[58] Yamaguchi, M., Nishina, N. (1995) Characterization of regucalcin effect on proteolytic activity in rat liver cytosol: relation to cysteinyl-proteases. *Mol. Cell. Biochem.*, 148:67-72.

[59] Mellgren, R. I. (1987) Calcium-dependent proteases: an enzyme system active at cellular membranes? *FASEB J.* 1:110-115.

[60] Croall, D. E., Demartino, G. N. (1991) Calcium-activated neutral protease (calpain) system: Structure, function, and regulation. *Physiol. Rev.* 71:813-847.

[61] Rosser, B. G., Powers, S. P., Gores, G. J. (1993) Calpain activity increases in hepatocytes following addition of ATP. Demonstration by a novel fluorescent approach. *J. Biol. Chem.* 268:23593-23600.

[62] Baba, T., Yamaguchi, M. (1999) Stimulatory effect of regucalcin on proteolytic activity in rat renal cortex cytosol: Invovement of thiol proteases. *Mol. Cell. Biochem.* 195: 87-92.

[63] Nishizuka, Y. (1986) Studies and perspectives of protein kinase C. *Science.* 233:305–312.

[64] Yamaguchi, M. (2013) Suppressive role of regucalcin in liver cell proliferation: Involvement in carcinogenesis. *Cell. Prolif.* 46:243-253.

Index

B

bacterial infection, viii, 24, 29
bacterium, 90
barriers, 113
Belgium, 23
beneficial effect, viii, 24
biochemical characterisation, vii, 2
biochemistry, 46, 87
biological systems, 10, 118, 131
biologically active compounds, 39
biosynthesis, 134, 156
blood, 5, 13, 20, 113, 115
blood-brain barrier, 13, 20
body weight, 107, 108, 109, 117, 130, 131
Bolivia, 90
bone, 6, 16, 19
bradykinin, 45
brain, 6, 13, 17, 20, 75, 158
breast cancer, 11, 19
bronchial epithelial cells, 31, 33, 37, 42, 49
bronchial epithelium, 33, 45, 48
bronchiectasis, 24
bronchitis, 43
butadiene, 130

C

calcitonin, vii, 108, 110
calcium channel blocker, 60
calcium oscillation, ix, 40, 67, 68, 69, 70,
    71, 72, 73, 74, 75, 76, 77, 79, 80, 82, 83,
    84
calcium-activated chloride currents, viii, 24
cancer, vii, 2, 11, 12, 20, 69, 72, 77, 80, 87
cancer cells, 11, 69, 72, 80, 87
candidates, 80
carboxyl, 41
carcinogen, 130
carcinogenesis, 11, 19, 155, 160
cardiac muscle, 9, 52
cascades, 26, 28, 37, 49
case study, 101
casein, 6, 7

cation, 25, 150
cattle, 90, 91, 100, 101
Caucasians, 24
CBP, 53, 55, 59, 61, 62, 63
cDNA, 92, 98, 133, 135, 156
cell biology, 40, 41, 46, 69
cell culture, 38
cell cycle, ix, 68, 69, 72, 73, 74, 76, 77, 78,
    79, 80, 83, 86, 87, 122, 123, 139
cell death, ix, x, 41, 68, 106, 107, 123, 124,
    125, 126, 127, 128, 129, 135, 136, 140,
    143, 146, 157
cell differentiation, 72
cell division, 11, 72, 74, 79, 86, 139
cell invasiveness, 12
cell killing, 146
cell line(s), 11, 12, 46, 47, 69, 77, 79, 87
cell metabolism, 131
cell signaling, 53, 73, 79, 86, 134, 156
cell surface, 15, 34, 35, 140
cellular regulation, vii, 132
cellulose, 146
cephalosporin, 109
cercaria, 98
cerebral cortex, 13
chaperone action, viii, 2, 14
chaperones, 20
chemical(s), 6, 35, 109, 110, 129, 130
chemotherapy, 42
CHO cells, 86
chondrocyte, 17, 18
cilia, 33, 39
circadian rhythm, ix, 68, 75, 76, 77, 86
circulation, 29
citrulline, 118
class switching, 43
cleavage, 74, 84, 140
clone, 135
cloning, 133, 155, 156
clustering, 27
clusters, 63
CNS, 20
cocoa, 13
coding, 98, 133, 156
coenzyme, 130

D

H

I

L

M

J

K

N

O

P

vitamin D, vii, 1, 2, 3, 4, 5, 6, 10, 11, 15, 17, 18, 19, 20

W

water, ix, 68, 71, 76, 77, 78, 79, 86, 87
water structure, ix, 68, 71, 77, 79
wavelet, viii, 51, 55, 56, 57, 59, 61, 64, 65
wavelet analysis, 56, 59, 64
Western blot, 5, 10, 120, 145, 146
wheat germ, 145
withdrawal, 125
wool, 90

X

X chromosome, x, 106, 144

Y

yeast, 64, 111
yield, 90

Z

zinc, 2, 8